Birmingham Archaeology Monograph Series 8

Life and Industry in the Suburbs of Roman Worcester

Edited by

Simon Butler and Richard Cuttler

Principal contributors

Robert Burrows, Simon Butler, Richard Cuttler, Lucie Dingwall, C. Jane Evans, Mike Napthan, and Eleanor Ramsey

with

Eleanor Blakelock, Angie Bolton, Robert Bracken, Marina Ciaraldi, Rowena Gale, Nick Garland, James Greig, Annette Hancocks, Emma Hancox, Rob Ixer, Erica Macey-Bracken, Helen Martin-Bacon, Jon Millward, Isabel Nogueira, Elizabeth Pearson, Stephanie Rátkai, Sian Reynolds, David Smith, Wendy Smith, Emma Tetlow, Jane Timby, Roger Tomlin, Roger White, David Williams, Felicity Wild, and Steven Willis

Illustrations by

Nigel Dodds, Helen Moulden, and Bryony Ryder

BAR British Series 533
2011

Published in 2016 by
BAR Publishing, Oxford

BAR British Series 533

Birmingham Archaeology Monograph Series 8
Life and Industry in the Suburbs of Roman Worcester

ISBN 978 1 4073 0800 5

BAR Publishing is the trading name of British Archaeological Reports (Oxford) Ltd. British Archaeological Reports was first incorporated in 1974 to publish the BAR Series, International and British. In 1992 Hadrian Books Ltd became part of the BAR group. This volume was originally published by Archaeopress in conjunction with British Archaeological Reports (Oxford) Ltd / Hadrian Books Ltd, the Series principal publisher, in 2011. This present volume is published by BAR Publishing, 2016.

Printed in England

BAR
PUBLISHING

BAR titles are available from:

BAR Publishing
122 Banbury Rd, Oxford, OX2 7BP, UK
EMAIL info@barpublishing.com
PHONE +44 (0)1865 310431
FAX +44 (0)1865 316916
www.barpublishing.com

CONTENTS

Figures

Chapter 1: Introduction

Chapter 2: St. Martin's Gate

Chapter 3: 1 The Butts

Chapter 4: 8–12 The Butts

Chapter 5: 14–24 The Butts

Plates

Tables

Chapter 2: St. Martin's Gate

Chapter 3: 1 The Butts

Chapter 4: The Butts

Chapter 5: 14–24 The Butts

CHAPTER 1: INTRODUCTION

Simon Butler

SUMMARY

Between June 2000 and April 2004 four sites within the City of Worcester were subjected to archaeological investigation by Birmingham Archaeology (formerly Birmingham University Field Archaeology Unit) and Mike Napthan Archaeology. The results from these four sites are documented in this volume. One site is located to the northeast of the historic city core at St Martin's Gate. The three remaining sites are located to the north of the city in an area known as The Butts. Archaeological excavations were undertaken at 8–12 and 14–24 The Butts, and an evaluation at 1 The Butts.

At St Martin's Gate the predominant features dating between the 1st and 3rd centuries were pits containing debris from iron smelting and smithing. These were sealed by a layer which indicated cultivation during the later Roman period. Later features also included a ditch, possibly part of the Civil War defences.

A single evaluation trench was excavated at 1 The Butts, and a watching brief observed during the redevelopment of the site. A Roman stone-lined well was identified, which had been backfilled in the later 4th century, and which contained stone, brick, and tile fragments derived from the demolition of one or more substantial buildings. Parts of the 13th-century city wall and ditch (filled in after the Civil War) were also identified.

Evaluation at 8–12 The Butts revealed a sequence of Roman, and later phases. The earliest occupation appears to date from the late 1st to early 2nd century, with a large pit sealed by gravel surfaces of similar date. Subsequent surfaces appear to be of 2nd- to 3rd-century date, with continued occupation of the area into the 4th century attested by residual material from post-medieval contexts. Later Roman or post-Roman 'dark earth' deposits survived close to the floor levels of a recently demolished building. Excavated post-medieval features included part of a possible Civil War ditch, and several quarry pits.

The earliest remains at 14–24 The Butts comprised a series of shallow pits and ditches, dated to the 2nd and early 3rd centuries. Large quantities of ironworking waste products were recovered, although no features associated with iron smelting were recorded. However, the waste products suggest this activity was taking place nearby. During the 4th century a hollow way was lined with a cobbled surface, the use of which possibly continued beyond the Roman period. It was sealed by a post-medieval deposit which contained large quantities of Roman finds. This may be the remains of a late/ post-Roman dark earth layer, or a late Roman midden, which was used to level the hollow way during the post-medieval period. Among the few other traces of post-Roman activity was a possible Civil War ditch. One other feature of note from the site was a well containing waterlogged plant remains indicating nearby pasture, or the margins of the occupied area.

BACKGROUND

This volume describes the results of archaeological fieldwork from four different sites within the City of Worcester (Figs 1.1 and 1.2). Three of the sites were located to the north of the historic city centre and one to the east; all lay immediately outside the former course of the medieval city walls. The archaeological works were undertaken in response to development projects which prompted evaluation and mitigation in accordance with PPG16. Due to the geographical proximity of the sites, and similarities in their character and chronology, it was decided to publish the results from all four projects as a single volume, in order to allow better comparison and synthesis of the archaeology.

Excavation and watching brief was undertaken in June 2000 at St Martin's Gate (Fig. 1.2), located to the northeast of Worcester, at the junction of St Martin's Gate and City Walls Road (NGR SO 8523 5499; see Chapter 2). The site lies immediately to the southeast of the extrapolated line of a former Roman road to Droitwich, in the proximity of a former medieval city gate. Trial trenching indicated the presence of Roman deposits, which led to further area excavations within the footprint of the new building, and a watching brief during localised ground works.

Whilst St Martin's Gate is to the northeast of the city, the remaining three sites in this volume are situated in close proximity to each other in an area known as The Butts (Fig. 1.3). This area takes its name from the road which follows a line along the exterior of the northern circuit of the city's former medieval defences. The place-name is often associated with medieval archery practice ranges, normally found on the perimeters of many medieval towns and cities. At 1 The Butts (NGR SO 8480 5410; see Chapter 3), situated on the southern side of the roadway, work within an archaeological evaluation trench cut across the medieval city ditch focused on a berm between the ditch and the medieval city wall.

The remaining two excavations were undertaken to the north of the road. At 8–12 The Butts (NGR SO 8478 5512; see Chapter 4), two evaluation trenches were excavated and revealed an unusually well-preserved sequence of Roman and immediately post-Roman deposits. Immediately to

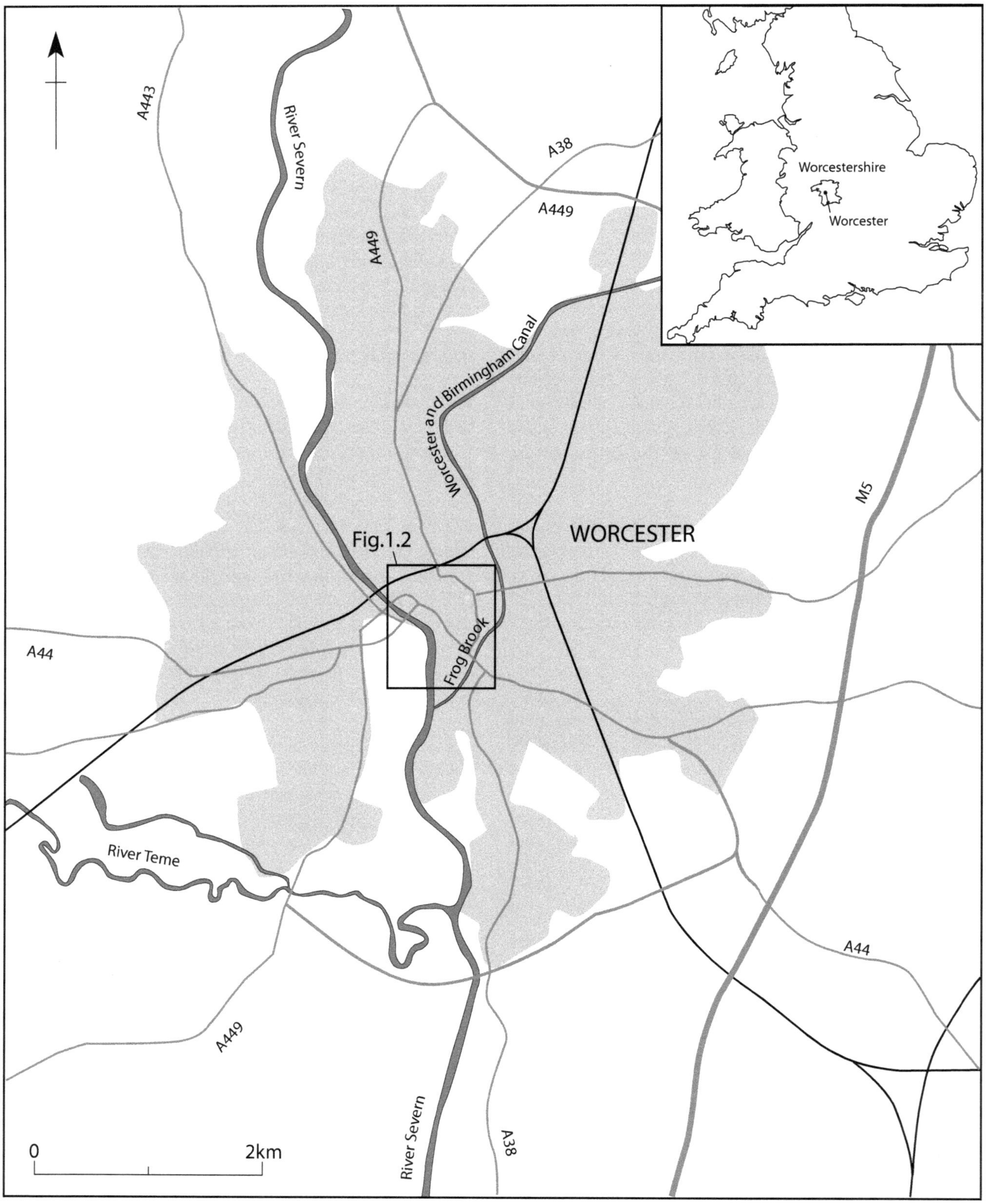

FIGURE 1.1 LOCATION OF WORCESTER

the west, at 14–24 The Butts (NGR SO 8473 5512; see Chapter 5), evaluation trenches in July 2000 demonstrated the presence of discrete Roman features. This led to the excavation of four areas within the footprint of new buildings between August and October 2003 and a subsequent watching brief maintained until April 2004.

THE ARCHAEOLOGY OF ROMAN WORCESTER

The historic centre of Worcester stands on the east bank of the River Severn on a terrace of sand and gravel, the Second (Worcester) Terrace, which forms a north–south ridge of well-drained higher ground above the river floodplain (Fig. 1.2). The underlying solid geology is Mercia Mudstone, which emerges at the eastern and western edges of the gravel terrace. Floodplain alluvium is found most extensively to the west of the Severn, but a narrower band also occurs on the eastern side; within the northwestern and eastern parts of the city centre. The narrow floodplain to the east of the city centre traces the former alignment of the Frog Brook, which would have been extant during the Roman period.

The soils on the Worcester terrace are light and well drained and the location of the Roman settlement appears ideal for defence. The location would also have controlled any traffic along the River Severn. While lithics ranging from the Palaeolithic to Bronze Age have been found within the city, there is as yet little stratified evidence for early prehistoric activity. Since remains of prehistoric date are likely to be ephemeral, and later activity extensive, their truncation by Roman and later activity cannot be ruled out. There is somewhat more evidence for late prehistoric occupation, including a possible Iron Age roundhouse at Deansway (Dalwood and Edwards 2004), but there is as yet scant evidence to elucidate the nature of pre-Roman Iron Age occupation in Worcester.

There remains a stark contrast between the slight evidence for prehistoric activity and the abundant evidence for Roman and later settlement, with evidence for Roman activity in Worcester more forthcoming. References from a 7th-century document listing all the towns of Roman Britain (The Ravenna Cosmography) suggest that the settlement was referred to as *Vertis*. The question arises of continuity of settlement from pre-Roman through to Roman times, and it may be that, despite the dearth of Iron Age evidence, the Roman settlement 'was an intensified, and by degrees urbanised, version of something that was already there' (Worcester City Council 2007, 22). Excavations by Barker (1968–69) at Lich Street (Fig. 1.2) provided a stratified sequence of probable defensive features extending from pre-Roman through Roman times, and Cunliffe (1991, 174) has suggested that Worcester could have been the site of a territorial *oppidum* of the Dobunni tribe, who inhabited this region in the Iron Age. Certainly, any pre-Roman settlement at Worcester would have been well placed to engage in the Dobunnic trade emanating from nearby Droitwich and Malvern, known respectively as important centres for salt and pottery production. It also seems likely that agriculture and metalworking would have characterised any Iron Age activity, as it did the Roman town.

The initial Roman interest in Worcester seems to have been based on military concerns, with the construction of an auxiliary fort, possibly for the campaigns into Wales. A defensive ditch at Lich Street (Fig. 1.2) and military artefacts at Deansway (Dalwood and Edwards 2004) and Sidbury (Darlington and Evans 1992), support the notion of a 1st-century fort but, beyond the evidence from Lich Street, the circuit and chronological development of any defences remains conjectural, and further evidence for the location of the fort is lacking. This could largely be a result of the distinct absence of archaeological work from the historic core of the city, as can be seen from Fig. 1.2. The River Severn was an important transport route in the early years of the Roman conquest, and although physical evidence for a bridge has so far proved elusive, there would almost certainly have been a crossing point on the Severn at Worcester.

It is clear that a Roman town was developing by the later 1st century. This is evidenced from a number of sites on the northern and eastern edges of the historic centre, which collectively suggest a rapid spread of domestic, agricultural and industrial activity over a wide but discontinuous area, spreading perhaps 1km from south to north along the sand and gravel terrace. The character of this early settlement is not entirely clear, but spatial variations are beginning to emerge. For example, at Sidbury, the evidence indicates fairly dense occupation with some large timber-framed 'strip buildings' of a type commonly used as shops or workshops in Roman small towns, and it has been suggested that Sidbury may have been a commercial market area of the town. In contrast, the evidence from Deansway for this early Roman period suggests lower-intensity occupation within agricultural plots or ditched compounds engaged in stock rearing and crop processing. At many sites, the occurrence of iron slag also points to the presence of an ironworking industry in the town, although this seems to have become more significant in the 2nd century.

Roman Worcester reached its apogee in the 2nd and 3rd centuries. A new and substantial ditch at Lich Street dates from some time after the 2nd century and is thought to have been part of a defensive circuit ditch enclosing the town centre (Fig. 1.2). Unfortunately, hard evidence for a Roman 'town centre' has yet to be found. Barker (1968–69) identified a suburb to the north of the ditch, where excavations such as those at Blackfriars, Broad Street, Farrier Street (Fig. 1.2), and Kardonia (Fig. 1.3), pointed to a possible industrial ironworking emphasis in this northern suburb. This may be an oversimplification, as more recent work suggests a wider distribution of ironworking throughout the Roman town and also the occurrence of other industries such as glass working and pottery manufacture. Jackson (2004, 102), identifies several ironworking foci and proposes shifts in focus through time during the 1st to 4th century. Thus, early

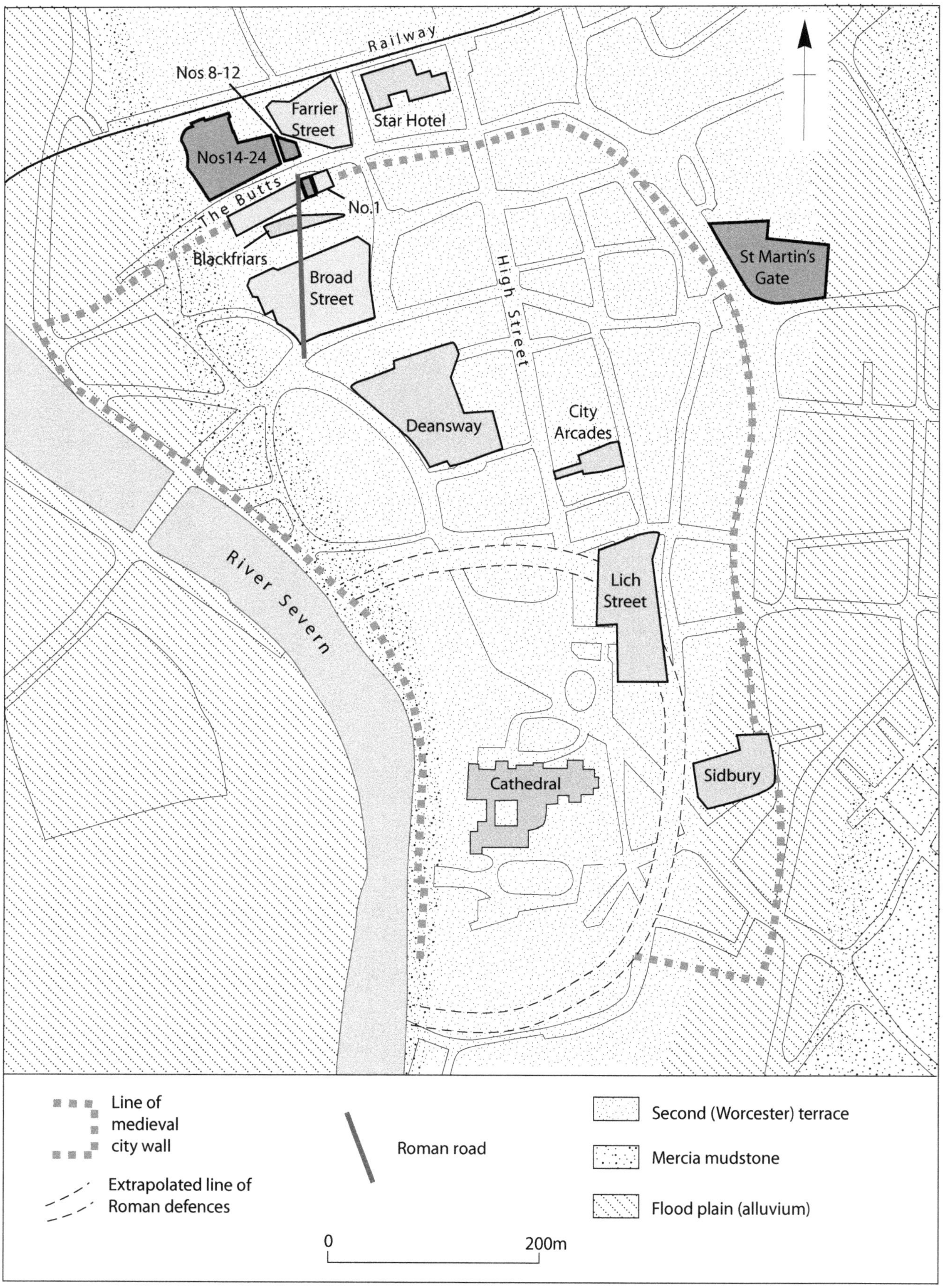

Figure 1.2 Location of sites and the geology of Worcester

activity is evidenced in the Deansway and Sidbury areas, followed by a shift to Broad Street and the northern suburbs during the late 3rd and 4th centuries. So far, the evidence suggests that iron smelting was more important than iron smithing, but both probably took place. Most analyses of the slag have pointed to an iron ore source in the Forest of Dean, some 40km southwest of Worcester, but more local sources cannot be ruled out.

The volume of Roman ironworking waste from Worcester led Burnham and Wacher (1990) to identify Worcester as a specialized ironworking centre in the context of Roman Britain. This idea may still hold true to some extent – Jackson (2004, 102) concludes that ironworking was certainly undertaken on a major industrial scale from the early 2nd century – but the Deansway excavations have also highlighted the role of other economic activities. In particular, animal husbandry intensified at the site during the 2nd and 3rd centuries, with an apparent emphasis on beef cattle. There is evidence that livestock were penned on the site, probably within hedged paddocks. Soil micromorphology (Macphail 2004) suggests the animals were grazed on the extensive adjacent floodplains of the River Severn. Dalwood and Edwards (2004, 47) state that 'the concentration of evidence for penning is interpreted as evidence for a degree of specialized livestock farming based on Worcester, and it is important to weigh this evidence against the more concrete evidence for iron production'.

It certainly appears that the 'northern suburb' of Roman Worcester included farmyards, and it should perhaps be seen as a mixed industrial, agricultural, commercial and residential area with a variety of buildings, yards and irregular enclosures. These were presumably connected by a network of lanes, minor roads and streets, which have been revealed in the form of cobbled and metalled surfaces from a number of excavations. However, the evidence is patchy, and the basic layout of roads through the settlement is not yet clear. The regional setting of Worcester points to at least two major roads entering the settlement; one from Gloucester to the southeast, which appears to have entered the town at Sidbury, and one from Droitwich to the northeast, which may have entered the town at Martin's Gate via Lowesmoor. Unfortunately, the line of the road from Droitwich has never been observed at Lowesmoor, which suggests that it may have skirted this once marshy area. Whilst the alignment of these roads can generally be traced in the countryside, their course is lost within the built-up area of today's city.

A third main road entering the town from Greensforge to the north has also been postulated on the basis of excavations at Broad Street, Blackfriars and Farrier Street, but other excavations along the projected line of this road at Kardonia and at the Magistrates's Court (not illustrated) have failed to verify its course. Further unsubstantiated roads have been postulated entering the town from the south and from the west. It is possible, but not yet proven, that the present High Street/ Foregate Street formed the main north–south axis through the Roman settlement. Within the northern suburb, excavations at Deansway have indicated that a regular layout of minor east–west roads was established in the mid 2nd century, perhaps linking the main north–south roads.

An uneven contraction of Roman Worcester during the 4th century is marked in the northern suburb by changes in land use and abandonment. At Deansway, for example, buildings that appear to have a primary industrial function were replaced with stockyards and a small cemetery, and roads fell into disrepair and were covered by rubbish and agricultural middens. By the late 4th or 5th century the general impression is that Worcester was a much smaller settlement with less emphasis on industry and with most of the suburbs given over to grazing.

RESEARCH THEMES

The current research themes for Roman Worcester are detailed in the city's urban archaeological strategy (Worcester City Council 2007). These themes provide a context for the archaeological projects in this volume, and they can be related to the four sites more specifically as follows:

Road network and settlement morphology

The four sites are situated in the 'northern suburb' of the Roman town, although St Martin's Gate is more to the northeast. To date, the northern extent of the city has been subjected to more extensive excavation than other parts Worcester, with the three sites at The Butts (Nos 1, 8–12, and 14–24; Figs 1.2 and 1.3) located close to sites previously excavated at Farrier Street (Dalwood *et al* 1994), Blackfriars (Barker 1968–69; Mundy 1986) and Broad Street (Barker 1968–69). Excavations at these sites revealed a north–south road (the postulated Greensforge road), the course of which might be expected to continue through 8–12 The Butts or 14–24 The Butts. The site at St Martin's Gate lies to the northeast of the hypothetical Roman core settlement, and very close to the postulated line of a southwest–northeast aligned Roman Road linking Worcester with Droitwich. Tracing the main road network through the Roman town is a research priority for Worcester, and the identification of any minor east–west roads, such as the three cobbled streets found at Deansway, would similarly add to an understanding of settlement layout and morphology. One of the questions that needs to be addressed is to what extent buildings or ribbon developments can be associated with the main roads, and to what extent there was simply an urban 'sprawl' from the main Roman centre. Similarly, there is a need to establish the general coherence of the settlement and its degree of focus on a currently unsubstantiated 'town centre', or consider the possibility of a more diffuse settlement with several foci.

Buildings and structures

A range of building types has been identified from Roman Worcester. Evidence for masonry structures is rather

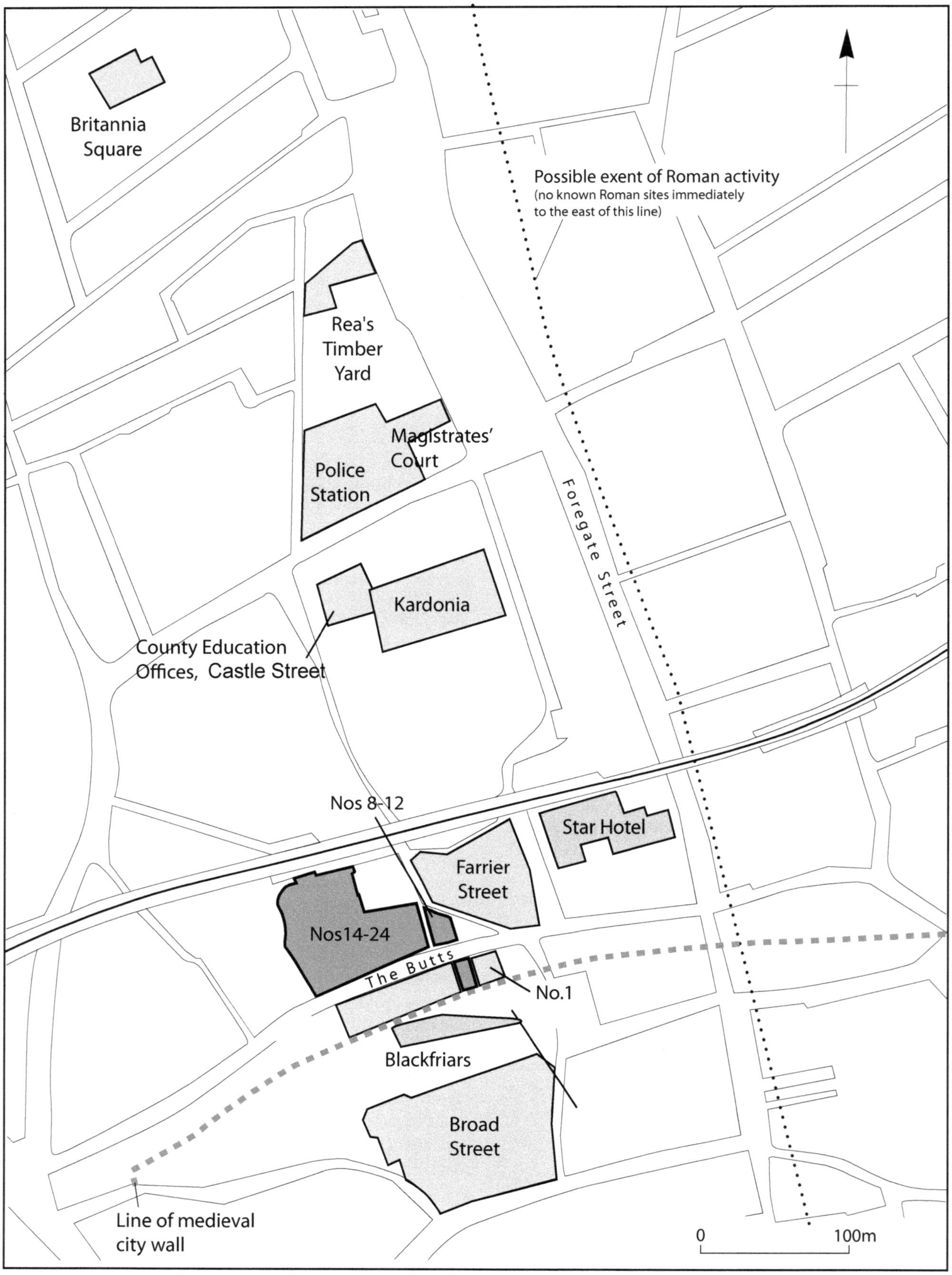

FIGURE 1.3 LOCATION OF EXCAVATED SITES CLOSE TO THE BUTTS

limited, particularly *in situ* stone foundations. However, these do occur at Britannia Square, 600m to the north of The Butts (Fig. 1.3), and at the Guildhall and City Arcades site on the High Street, some 400m southeast of The Butts. It is uncertain if these foundations relate to civic public buildings or to wealthy private buildings. A good quantity and range of dumped building material, including roof tiles, brick, painted wall plaster and window glass, found at Sidbury indicates traditional Roman buildings, although their locations are unknown. At Broad Street only two fragments of painted wall plaster were recovered, while other sites in the northern suburb, including Farrier Street, Blackfriars and Deansway also lack evidence for substantial Roman buildings. Indeed, the evidence so far from the northern suburb suggests mostly timber-built structures, such as those found in association with farmyard enclosures at Deansway. Any new structural evidence from the northern suburb therefore, would contribute to an understanding of townscape and land-use patterns.

ECONOMY

Ironworking was undoubtedly a major industry in Roman Worcester, with production sites so far excavated at Broad Street and Deansway, however, the role of other economic activities should not be underestimated. These include stock rearing and glass production, as recorded at Deansway, and pottery manufacture. Discoveries such as those at Deansway call into question the concept of suburbs specialising in ironworking, and emphasise the need for further evidence relevant to understanding the actual nature and development of the Roman town's economy, and how it may have related to wider economic networks and social circumstances.

LATE ROMAN AND POST-ROMAN 'DARK EARTH' DEPOSITS

Across Britain, many archaeological deposits from the late Roman/ post-Roman period are characterised by the formation of a dark, organic-rich soil which is commonly termed 'dark earth'. These generally reflect a poorly stratified layer overlying Roman deposits with scant evidence for *in situ* features. Its origin has been variously interpreted as the result of agricultural or horticultural activity, waste disposal or land abandonment. In the context of Worcester, and more specifically Deansway, Macphail (2004, 78) suggests that this material 'represents the formation of a rural grassland landscape after late Roman abandonment of open settlement areas'. More importantly, he warns that what are perceived as similar dark earths on different sites may in reality have very varied origins, thus similar depositional processes cannot be assumed even within the context of adjacent excavations. At Deansway, Macphail's soil micromorphology suggested that the layer resulted from use of the site as grazing land between the 4th and 9th centuries. This information has been important in documenting late Roman and early post-Roman land-use change and settlement contraction in Worcester, and it seems that cultivation or 'dark earth' deposits are the main source of material from this period. Further examination of these deposits is therefore a research priority for the city.

ARRANGEMENT OF THE PUBLICATION

This volume details the results of the fieldwork investigations in Chapters 2–5. Each chapter introduces the site background, methodology, with a summary of the archaeological deposits followed by specialist reports. Chapter 2 is concerned with the excavation at St Martin's Gate. Chapter 3 describes the evaluation at 1 The Butts. Chapters 4 and 5 describe the fieldwork at 8–12 The Butts and 14–24 The Butts, respectively. Chapter 6 provides a synthesis of the results with a summary of the dates, phases and main features.

CHAPTER 2: ARCHAEOLOGICAL FIELDWORK AT ST MARTIN'S GATE

Lucie Dingwall and Eleanor Ramsey

with Marina Ciaraldi, C Jane Evans, Rowena Gale, James Greig, Annette Hancocks, Stephanie Rátkai, Roger White, and Steven Willis

SUMMARY

The site at St. Martin's Gate had been heavily disturbed in the post-medieval period, but Roman remains were discovered in discrete areas of the site, dating mainly between the 1st and 3rd centuries (Phase 1). The remains consisted primarily of a series of pits, the contents of which indicated that iron smelting and smithing had been carried out at, or very close to, the site in the Roman period. The Phase 1 remains were sealed by a cultivation layer, deposited during, or after the late 3rd to 4th century (Phase 2). No intact medieval remains were discovered in the excavated areas, but a ditch of probable Civil War origin was aligned north–south across the centre of the site.

INTRODUCTION

This chapter describes the results of an archaeological evaluation, watching brief and area excavation carried out in June 2000 on a site at the junction of St Martin's Gate and City Walls Road, Worcester (NGR SO 8523 5499; Fig. 1.2). The site comprised a parcel of land formerly occupied by a car showroom and garage, but cleared of buildings in 1998–99, prior to construction of a new car showroom and associated external works. Clearance included the removal of substantial underground fuel tanks, with no structures remaining at the time of the archaeological work, apart from boundary walls and fencing.

The site lies within the designated historic core of the Roman and medieval city (Archaeologically Sensitive Area no. 18), and there was a possibility that groundworks associated with development might disturb archaeological remains. As the groundworks consisted of vibro-piling within the footprint of the new building, and more shallow disturbances for drainage, services and boundary walls around the new building, a two-stage approach to the archaeological recording was adopted. This comprised a first stage of evaluation by trial trenching, with extensive excavation within the footprint of the new building prior to construction, and a second stage of watching brief during construction. The watching brief included drainage and service trenches, boundary walls, security posts and bollards and other associated works. In the areas that had previously been disturbed by the removal of underground fuel tanks, archaeological monitoring was not deemed necessary.

The trial-trenching phase indicated that significant deposits from the Roman period had survived in discrete areas of the site. The objective of archaeological recording was to provide a detailed record of the character, extent and significance of any such remains to be disturbed by the development, in order to contribute towards a greater understanding of this area of Roman Worcester.

SITE BACKGROUND

The site at St Martin's Gate lies to the northeast of the historic centre of Worcester and the postulated core of Roman settlement. It is also immediately to the southeast of the postulated line of a southwest–northeast aligned Roman road linking Worcester with Droitwich (Fig. 1.2), and it has been suggested that Roman settlement may have extended some distance along such roads (Barker 1968–69). The site also lies close to the former course of the Frog Brook (Fig. 1.1), the nearby presence of a water course being ideal for industrial activities, such as ironworking (Carver 1980a, 26). Previous archaeological observations to the east of the historic city have taken place at Tallow Hill/ Pheasant Street (Whitworth and Edwards 2001) and at Spring Gardens (Mundy 1994), but these revealed little direct evidence for Roman activity.

The Frog Brook (Fig. 1.1) was diverted into the town ditch in the medieval period, and part of the area around the brook was known to be marshy in the medieval period (Carver 1980a, 21). The site is located just outside the medieval city walls, within the medieval suburb of St Martin's Gate/ Silver Street, just outside the St Martin's Gate entrance to the historic city. The medieval St Martin's Gate was the last of the ancient gates of the city to survive, being demolished in the 18th century, and is described as having had a 'hornwork of defence' (VCH 1924, 377). Little is known from the archaeological record about the land outside St Martin's Gate. It might be expected that the area immediately outside the city walls would have been the domain of industrial activities which could not be located within the confines of the city, and that it would have been occupied by the dwellings of the lower social groups which are notably absent from the archaeological record (White and Baker 2000). Documentary sources suggest a different picture, however, although the information is perhaps slightly contradictory. The area between Silver Street and the city wall was apparently still undeveloped in 1630 when 'William Beauchamp rented a "little messuage" there and a meadow adjoining with three racks' (Hughes 1980, 286), but other sources indicate that relatively wealthy citizens had houses in the suburb in the first part of the 17th century. However, the sources also indicate that almost all of the houses outside St Martin's Gate were burnt down and destroyed in the Civil War, although it appears that the building that became the Plough Inn (demolished in the 1970s) survived the Civil War demolition (Hughes 1980, 286).

A stylised 1651 plan of the city depicts the area immediately outside St Martin's Gate as cultivated fields, with a Civil War bastion shown defending the gate. Doharty's plan of 1741, a more reliable source, shows buildings all along the street frontages of the site, and these are also shown on Broad's plan of 1768, but no structures further back from the frontages. Young's map of 1779 and Nash's map of 1781 show buildings covering most of the site, and Valentine Green's map of 1795 depicts the buildings in more detail, marking a vinegar manufactory in the northeastern area of the site. Stratford's plan of 1835 and Dewhurst and Nichol's plan of 1838 show less detail, but the 1888 First Edition Ordnance Survey map shows a tannery occupying the northeastern part of the site, the vinegar works having shifted to the east. By this time, the rest of the site was occupied by housing, with some open areas retained as courtyards.

METHODOLOGY

An L-shaped trial trench, approximately 15m x 15m long and 1.6m wide, was excavated within the footprint of the new building (Fig. 2.1, Area 1). This was located along the northwestern boundary of the building footprint in order to avoid an area disturbed by underground fuel tanks. The trench revealed remains dating to the Roman period, and,

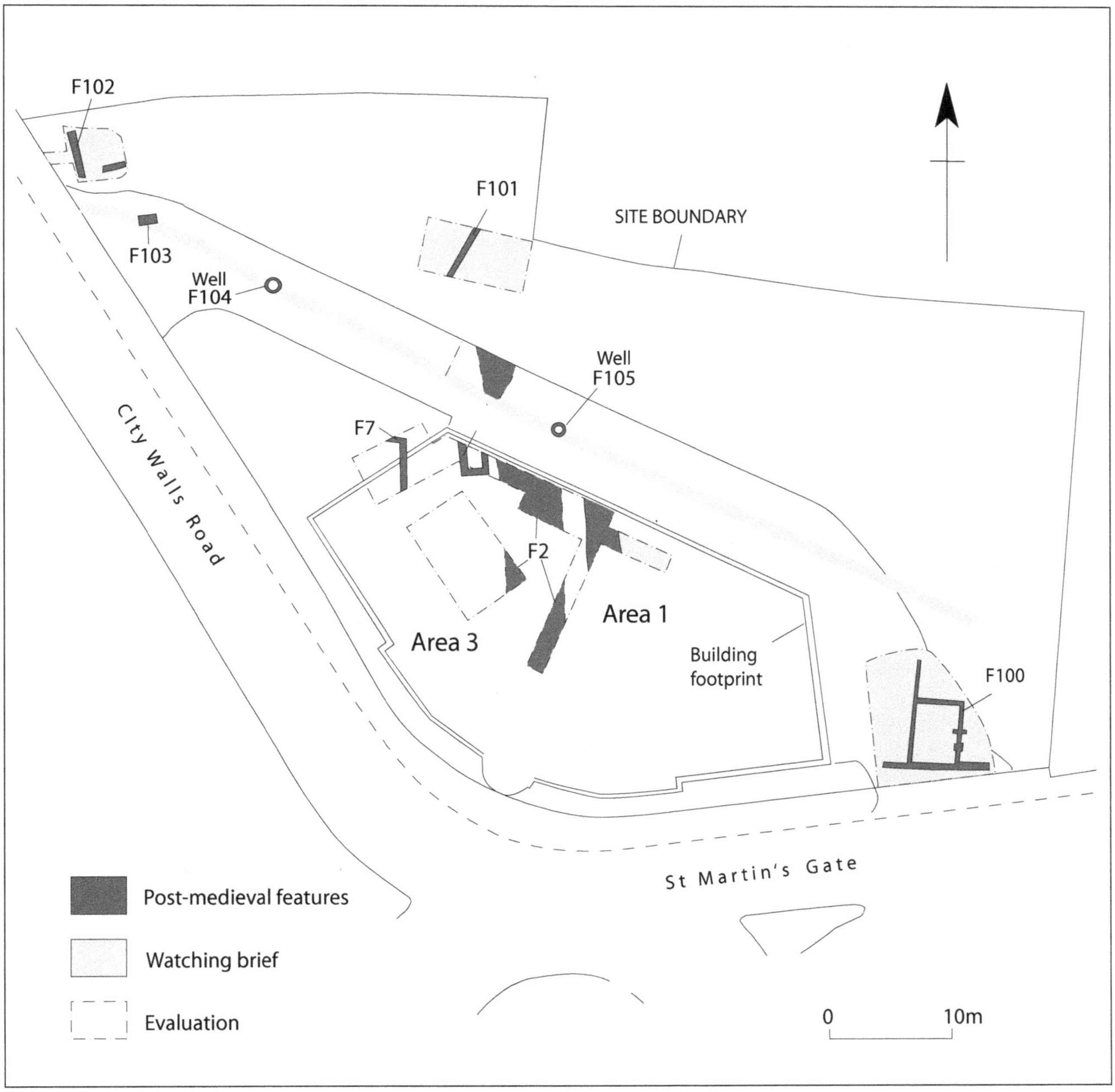

FIGURE 2.1 GENERAL SITE PLAN

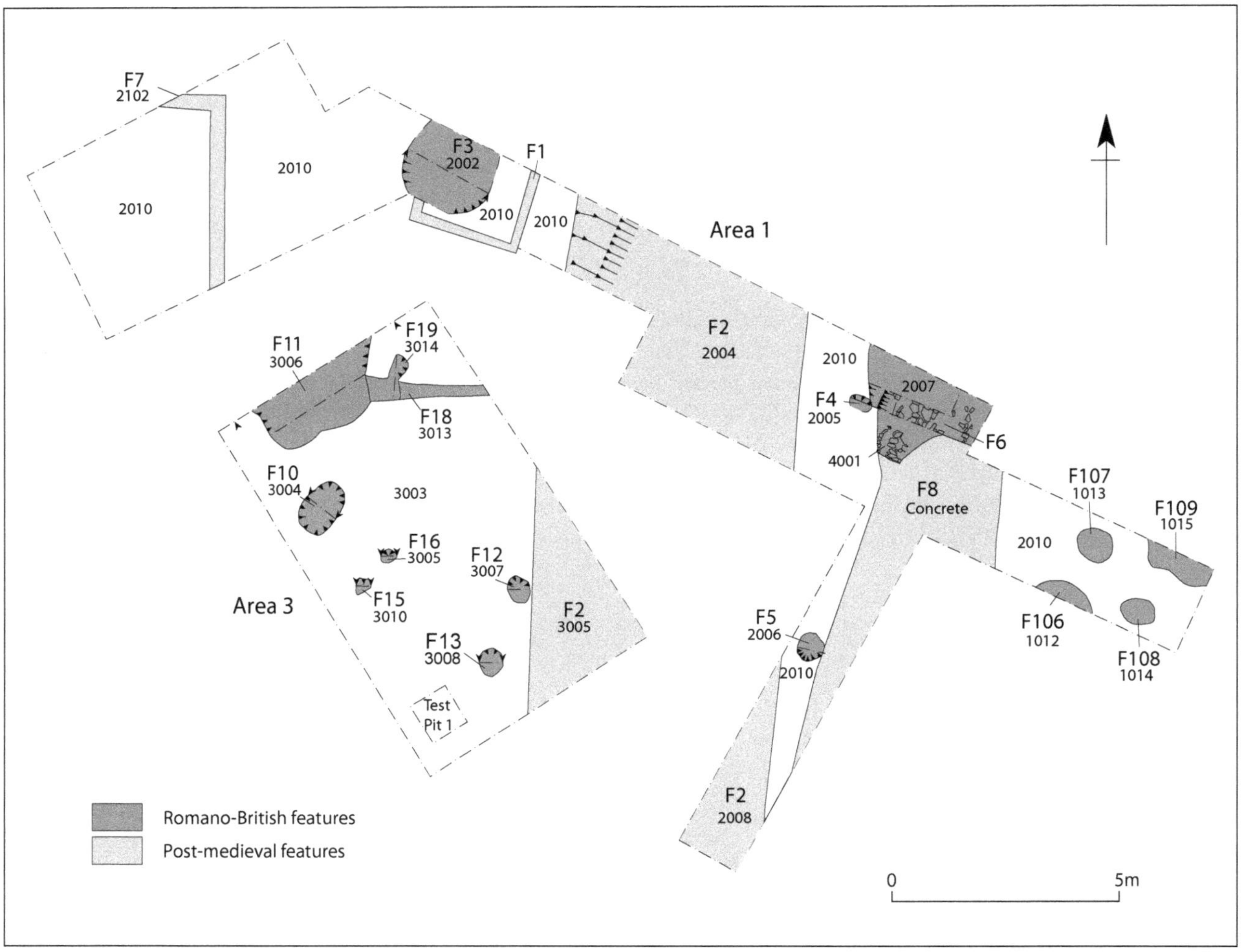

FIGURE 2.2 PLAN OF AREAS 1 AND 3

subsequently, more extensive excavation was undertaken in areas of surviving remains. The trench was extended to the northwest, widening it at the eastern end (Fig. 2.2), and excavating a new area, approximately 6m x 10m in size, to the south of the original trench (Fig. 2.2, Area 3).

RESULTS

On the basis of the pottery (see Evans, this chapter) activity on the site can be divided into three phases. Phase 1 dates to the late 1st to early 3rd century. Phase 2 is represented by a cultivation layer, which on the basis of the ceramic evidence, could have started accumulating as early as the late 3rd century. Similar cultivation deposits from other sites in Worcester have been attributed to the late Roman and post-Roman periods. Phase 3 dates to the 17th to 19th century, with some residual medieval material present.

PHASE 1: LATE 1ST TO EARLY 3RD CENTURY

AREA 1 (Figs 2.2–2.3)

In Area 1, the subsoil (2010) was encountered at a depth of approximately 1.8m below ground level, and was cut by three sub-circular pits (F3–F5) which produced Roman pottery. One of these (F3) was relatively large (2.0m in diameter and 0.65m deep), with steep sides and an irregular base, and had two distinct fills of green/ grey silty sand (2013 and 2002), which contained 54 sherds of Roman pottery, as well as charcoal, tile and slag. Further sub-circular pits (F106–F109) were recorded at the eastern extent of Area 1 during the watching brief. These could not be hand excavated for safety reasons, but their character, stratigraphy and relative depth suggested that they were contemporary with the Roman features to the west.

A possible north–south aligned linear feature identified in Area 1 (F6) could not be confirmed as such because it extended beyond the eastern limit of the excavation and was truncated to the south by a modern concrete foundation (F8). The feature (F6) was at least 2.75m wide and 0.5m deep, with a steep slope on the west side. Large limestone blocks were found at the base of the cut, some with signs of having been dressed or worked, and it is possible that this feature represents the cut for a wall. The fill (2007/ 4001), a brown clay silt, contained a small quantity of Roman pottery, a few cultivated cereal grains, and ironworking remains in the form of slag and spheroidal hammerscale.

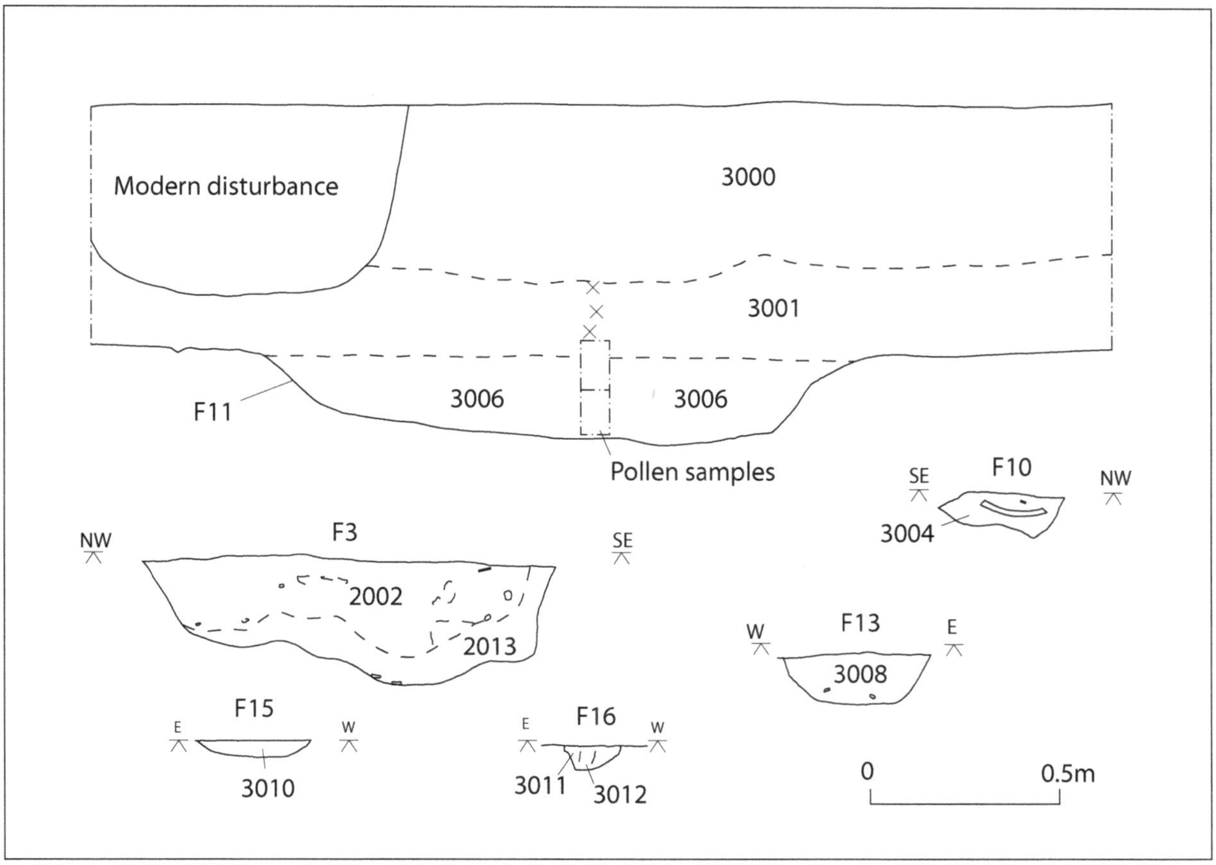

FIGURE 2.3 SELECTED SECTIONS FROM AREAS 1 AND 3

AREA 3 (Figs 2.2–2.3)

The subsoil (3003) was cut by several pits (F10–F13 and F15), which were sub-circular and ranged in size from 0.6m to 3m in diameter, and from 0.1m–0.5m in depth. A posthole (F16), which contained no datable material, was also recorded. All of the pits, except F15, contained Roman pottery. One of the pits (F10) contained large amounts of amphorae, as well as other Roman pottery, fragments of heat-cracked stone, spheroidal hammerscale, charred cereal grains and other crop-processing by-products (Ciaraldi, this chapter). Heat-related changes in the subsoil surrounding this feature indicated burning *in situ*, and large pieces of charcoal with ironworking slag still attached were recovered from the fill. Another of these pits (F13) contained fragments of vitrified charcoal, indicating that the charcoal had been heated to very high temperatures (Gale, this chapter).

The largest pit (F11), which was not fully exposed at excavation, had curving sides to a flat base, and a single fill (3006) that contained 98 sherds of Roman pottery and fragments of degraded bone. This pit cut an east–west aligned shallow feature (F18), possibly a beam-slot, which contained a few sherds of Roman pottery. Adjacent to the feature was a shallow scoop, possibly a truncated pit (F19), which also contained Roman pottery and fragments of oyster shell.

PHASE 2: LATE 3RD CENTURY ONWARDS (AREAS 1 AND 3)

All of the Roman features in Areas 1 and 3 were sealed by a very dark, grey-brown organic layer (3001, Fig. 2.3), 0.3–0.5m deep containing iron slag and Roman pottery, which ranged in date from the 1st to 3rd century, apart from one sherd of late 3rd- to 4th-century Black Burnished ware (see Evans, this chapter). The presence of the latter provides a *terminus post quem* of the late 3rd century for the layer, and indicates that some or all of the artefacts from the layer may have originated from elsewhere in the town. A copper alloy brooch dating to the late 1st century (see White, this chapter) was recovered from a Test Pit excavated through this layer in Area 3 (Fig. 2.2). Two pieces of post-medieval blast furnace slag were recovered from one of the test pits but, as no other post-Roman finds were recovered from this layer, it is likely that the slag was intrusive, and the layer is interpreted as a late Roman cultivation layer. However, as the layer was not completely hand-excavated, this interpretation is not conclusive.

PHASE 3: 17TH TO 19TH CENTURY (AREAS 1 AND 3)

Within the centre of Area 1, and the eastern extent of Area 3 (Fig. 2.2), was a large north–south aligned ditch (F2), approximately 5.5m wide. While the upper profile of the ditch was difficult to define due to later disturbance, the feature was not fully excavated for safety reasons.

Sections excavated through the lower part revealed it to be steep-sided, with a rubble fill (2004, 2008, and 3005) containing pottery ranging in date from the 17th to 19th century. Finds included part of a two-handled, blackware cup or mug from the mid to late 17th century, and brick, tile and mortar. Sondages excavated through the ditch fill demonstrated that the feature reached a maximum depth of 3m below the modern ground surface and had completely removed earlier deposits. Demolition rubble (2010) sealed both this ditch and a post-medieval cellar wall in the west of Area 1 (F7, Fig. 2.2). The rubble was cut by three sides of a small brick structure (F1), to a depth of 0.7m below the surface, and by a modern disturbance (F8), containing concrete piles, which had truncated earlier features (Fig. 2.2).

During the watching brief, various other post-medieval walls were recorded at an average depth of 0.5m below the modern ground surface, including a probable cellar abutting parts of a larger building in the southeast of the site (F100, Fig. 2.1), and a northeast–southwest aligned brick wall in the north of the site (F101, Fig. 2.1). In the northwest corner of the site, walls constructed of hand-clamped bricks, set on a foundation consisting of two courses of sandstone blocks, were recorded (F102 and F103, Fig. 2.1). Dense deposits of post-medieval tile were dumped against these walls, which were cutting post-medieval ground make-up and were sealed by modern brick and concrete layers. Two post-medieval brick wells (F104 and F105, Fig. 2.1) were also identified in section during the excavation of a drainage trench. All features were overlain by post-medieval rubble and levelling layers with a combined depth of 1.8m in Area 1, and 1m in Area 3.

FINDS

ROMAN BROOCH *Roger White*

INTRODUCTION

A Roman brooch was recovered from a test pit (Test Pit 1, Fig. 2.2) excavated through the cultivation layer in Area 3 (3001). It was stabilised through conservation at Wiltshire Museums Service. The conservation involved mechanical removal of corrosion products with the cleaned surface being coated with Incralac.

DESCRIPTION

The overall dimensions of the brooch were *c* 60mm by 23mm. Despite conservation, the brooch is in a poor state of preservation, with the original surface only surviving on the dorsal and ventral parts of the brooch. However, conservation has at least arrested further decomposition. In design, the brooch is basically T-shaped, with a strongly arched bow. The arms are cylindrical in section but are poorly preserved, especially at their ends, so that only the axle bar is prominent here. The only decoration visible is two dots placed one above the other at the junction of the arms and the bow; those on the left hand side are better preserved and one still holds a dot of red enamel. A plain hook above the arms holds a remnant of the wire spring and below this, at the start of the bow, is a stud that has been identified in the conservation report as made of iron. A pierced hole lies below this and it too presumably once held a stud. The remainder of the bow has five panels that gradually decrease in area but which once held enamel. The second panel from the top is still filled with enamel that is now a dull grey colour but which may originally have been yellow or white. The end of the bow tapers away and is flattened for the catch plate which is, however, entirely lost.

DISCUSSION

The brooch can be confidently assigned to the Head-Stud brooch group; a large and important class of T-shaped brooches that evolved in the later 1st century. The group was in production and use until the later 2nd century, with some remaining in use perhaps into the 3rd century. The brooch was popular in the Midlands and there is strong evidence for regional types indicating localised production (eg, Philpott 1999). In genesis, the type is clearly derived from the Colchester brooches, so that the stud was originally conceived as a functional device for riveting the hook holding the brooch spring to the brooch bow (Hattatt 1989, 80–81, no. 1524). However, in the majority of Head-Stud brooches, the stud has achieved a purely skeumorphic and decorative function and is thus often highly decorated and/ or enamelled. Analysis of the pin mechanisms shows that the type could have a Colchester type sprung pin, a simple sprung pin or, later, a simple hinged pin. Our example belongs in the first group and thus is probably late Flavian in date (Mackreth 1994, 164–65).

Three decorative types of enamelling are known: a single cell running down the bow, a series of rectangular cells also on the bow, or a lozenge and triangle pattern that can be found on the arms in addition to the bow (Mackreth 1994, fig. 77, 36). This last type of enamelling is virtually exclusively reserved for the later hinged-pin variety and at least one example of this type is known from Worcester itself (Mackreth 1992, 73–75). The twin dots of enamel seen on the St Martin's Gate example are a rare feature and the author knows of no parallel for it. However, this is unlikely to be an important diagnostic feature and merely reflects the wide variety of producers of brooches of all types at this date (*ibid*).

ROMAN POTTERY *C Jane Evans with Steven Willis*

INTRODUCTION

A total of 432 sherds of Roman pottery, weighing approximately 6.5kg, was recovered during the excavation. Most came from Area 3 (Table 2.1), the largest groups being from two pits (F11, 98 sherds and F10, 62 sherds). The largest Area 1 assemblage also came from a pit (F3), which produced 54 sherds. All but two sherds came from

Area	Qty	% Qty	Wt (g)	% Wt	Av Sherd Wt (g)	Rim EVE	% Rim EVE
1	73	17	691	11	9.5	32	10
3	359	83	5757	89	16	279	90
TOTAL	**432**		**6448**		**15**	**311**	

TABLE 2.1 SUMMARY OF THE ROMAN POTTERY BY AREA

Phase	Qty	% Qty	Wt (g)	% Wt	Av. Sherd Wt (g)	Rim EVE	% Rim EVE
1	267	61.5	5251	81.5	20	193	62
2	163	38	1103	17	6.7	114	37
3	2	<1	94	1.5	47	4	1
TOTAL	**432**		**6448**		**15**	**311**	

TABLE 2.2 SUMMARY OF THE ROMAN POTTERY BY PHASE

WHEAS Fabric Code	Common Name	Archive Code	Description/ references (T&D = Tomber and Dore 1998)
3	Malvernian group A, handmade	N02.1	T&D MAL REA, 147, plate 120; Peacock 1965–67
8?	Black-surfaced grey wares	G08.6	See Ariconium (Willis forthcoming)
8?		G08.7	
8?		G08.8	
12	Severn Valley ware	O02.1	Standard oxidised fabric, unsourced: T&D SVW OX 2, 149, Pl 122; Webster 1976, Rawes 1982
12/ 20?		W16	Standard oxidised fabric with white slip
12.1		G04.1	Standard fabric, reduced
12.1/ 20?		W16.1	Standard fabric, reduced with white slip: misfired?
12.2		O03.1	Organic tempered variant, oxidised (elongated voids appearing as black/ dark grey streaks in fracture.
12.3		G05	Organic tempered variant, reduced
12.4		O02.2	Variant with limestone temper
13	Sandy oxidised wares	O06.18	
14	Sandy reduced wares	G00	
14/ 15		G02.2/ 2.3	Pale sandy grey ware with blackened surfaces (?LNVGW). Wheelmade, hard. Reduced throughout, with a light grey core (5Y 7/ 1) and dark grey surfaces (GLEY1 4/ N). Common white sand <0.1mm and other sparse iron rich inclusions.
16.2	Grog-tempered ware	F017	Cf Gloucester grog-tempered native ware (Ireland 1983 TF2, 100)
19	Malvernian group A, wheelmade		
20?	White-slipped sandy ware	W17	
22	Southeast Dorset BB1	B02	T&D DOR BB 1, 127, pl 100; Williams 1977; Seager Smith and Davies, 1993
41.1	Cream wares	P00	
42.1	Amphorae, Dressel 20	A01	Baetican Dressel 20, Peacock and Williams 1986, class 25; T&D BAT AM 1/ 2, 84–85, pl 61, 62
42.6?	Amphorae	A0	
43	Samian	S0	
109	Eggshell ware	P07.2	
34	'Wroxeter' mortaria	M015	White mortaria; T&D WRX WH, 179, pl 150a–c

TABLE 2.3 LIST OF FABRICS REPRESENTED

Common Name (Archive code)	WHEAS Fabric Code	Qty	% Qty	Wt	% Wt	Rim EVE	% Rim EVE
Severn Valley ware, oxidised (O02.1)	12	161	37	1367	21	88	28
White-slipped variant, oxidised (W16)	12/ 20?	3	<1	12	<1	0	0
Reduced variant (G04.1)	12.1	42	10	555	9	97	31
White-slipped variant, reduced (W16.1)	12.1/ 20?	3	<1	41	<1	0	0
Organic variant, oxidised (O03.1)	12.2	63	14.5	996	15	23	7.5
Organic variant, reduced (G05)	12.3	13	3	105	2	12	4
Limestone-tempered variant (O02.2)	12.4	8	2	71	1	0	0
Total Severn Valley ware		*293*	*67.5*	*3147*	*49*	*220*	*70.5*
Malvernian metamorphic (H)	3	21	5	156	2.4	3	1
Malvernian metamorphic (W)	19	6	1	131	2	0	0
Black-surfaced grey ware - sand (G08.6)	8?	6	1	11	<1	8	2.5
- sand and limestone (G08.7)	8?	1	<1	1	<1	0	0
- organic (G08.8)	8?	1	<1	3	<1	0	0
Sandy oxidised ware (O06.18)	13	1	<1	7	<1	0	0
White-slipped sandy oxidised ware (W17)	20?	1	<1	3	<1	0	0
Sandy reduced ware (G00))	14	1	<1	4	<1	0	0
Grog-tempered ware (F017)	16.2	10	2	159	2.5	0	0
Total Regional/ Local wares		**341**	**78.5**	**3622**	**56**	**231**	**74**
Southeast Dorset BB1 (B02)	22	33	8	240	4	29	9.3
Nene-Valley-type sandy reduced ware (G02.2)	14/ 15	1	<1	4	<1	2	<1
Nene-Valley-type sandy reduced ware (G02.3)	14/ 15	5	1	27	<1	11	3.5
Cream ware (P00)	41.1	2	<1	3	<1	0	0
Eggshell ware (P07.2)	109	1	<1	1	<1	0	0
Wroxeter mortaria (M015)	34	3	<1	33	<1	8	2.6
Total Traded ware		**45**	**10**	**308**	**5**	**50**	**16**
Miscellaneous		16	4	13	<1	0	0
Total uncertain source		**16**	**4**	**13**	**<1**	**0**	**0**
Amphorae, Class 25 (A01)	42.1	18	4	2418	37	0	0
Amphorae (A0)	42.6?	2	<1	58	<1	0	0
Total amphorae		*20*	*4.5*	*2476*	*38*	*0*	*0*
Samian, (S0)	43	12	3	39	<1	30	10
Total samian		*12*	*3*	*39*	*<1*	*30*	*10*
Total Imported		**32**	**7.5**	**2515**	**39**	**30**	**10**
TOTAL POTTERY		**434**	**100**	**6458**	**100**	**311**	**100**

TABLE 2.4 SUMMARY OF THE ROMAN POTTERY ASSEMBLAGE BY FABRIC/ SOURCE

stratified Roman deposits (Tables 2.1 and 2.2). The pottery was fairly fragmentary, but several diagnostic form sherds were noted (described below). These, together with the range of fabrics represented (Tables 2.3 and 2.4), indicated a predominantly 1st- to 2nd-century date for the Roman activity. There was evidence for some 2nd- to 3rd-century activity, and slight evidence for late 3rd-to 4th-century activity in the vicinity, but nothing diagnostically later.

METHODOLOGY

The pottery was recorded in full using the standard (BUFAU) Roman pottery recording system, described in the project archive. Fabrics were recorded with reference to the Worcestershire County Fabric Series, prefixed WCM in the text (Hurst and Rees 1992, 200–209). Precise form types and broad vessel classes (for example bowl, flagon, and mortarium) were both recorded. These were cross-referenced with illustrated forms from other Worcester sites (Darlington and Evans 1992; Bryant and Evans 2004) or other key publications. Other characteristics studied included decoration, evidence for manufacture (overfiring), repair (none noted), use or reuse (eg, counters). The assemblage was quantified by sherd count, weight and EVE (estimated vessel equivalent). Quantification by rim EVE is presented in this report, although both rim and base EVEs are recorded in the archive. The condition of the pottery varied, but all sherds were abraded to some degree. None of the pottery is illustrated, as all forms are well published elsewhere.

DISCUSSION

The best evidence for dating Phase 1 was provided by sherds of samian ware and Black Burnished ware (BB1). The samian ware indicated a date of *c* AD 120–40/ 50 and was found in features F3 and F11, although some residual 1st- and earlier 2nd-century material was also present. This date was broadly supported by the presence of BB1. This generally indicates a *terminus post quem* of *c* AD 120, although small quantities have been found in earlier contexts elsewhere in the region. The large pit in Area 3 (F11) was the only feature to produce form sherds in BB1. These comprised a flat-rimmed bowl or dish (Seager Smith and Davies 1993, fig. 123, type 22/ 3), a beaded rim beaker (*ibid*, fig. 123, type 10), a groove-rimmed bowl (*ibid*, fig. 123, type 24) and a dish with a slightly beaded rim (*ibid*, fig. 123, type 20). The first two of these date broadly to the 2nd century. The latter two, however, support a *terminus post quem* sometime in the later 2nd or earlier 3rd century, providing the latest dating evidence for this phase. This is consistent with some of the associated Severn Valley ware forms, for example a moderately splayed tankard and a segmental bowl (Webster 1976, fig. 7, 40; fig. 10, 65).

A high level of residuality was evident in the assemblage, and a significant quantity of 1st- to early 2nd-century pottery was included. Typically early coarseware included organic-tempered Severn Valley ware (WCM fabrics 12.2 and 12.3), fragments from a large, grog tempered storage jar (WCM fabric 16.2), Malvernian wares (WCM fabrics 3 and 19), black surfaced grey wares (WCM fabric 8), and white wares and white slipped wares (WCM fabrics 20, 41.1 and 109). The black surfaced grey wares are similar to the range of transitional fabrics noted at *Ariconium* (Willis 2000, 85). These, and the other fabrics, are all found in early Roman contexts elsewhere in Worcester (Darlington and Evans 1992, 36–39; Bryant and Evans, 2004). Some forms were also characteristic 1st- or 2nd-century types. These comprised a Severn Valley ware tankard and carinated bowl (Webster 1976, fig.7, 38; fig. 9, 59–60), and, in reduced Severn Valley ware, a 'waisted' bowl (cf. Darlington and Evans 1992, fig. 22, 7), fragmentary rusticated jars (cf. *ibid*, fig. 22, 1), and various beakers (cf. *ibid*, fig. 17, 1–3). Malvernian tubby cooking pots were also present (Peacock 1965–67, fig. 1, 1–8). Less well-dated fabrics included oxidised and reduced Severn Valley wares (WCM fabrics 12 and 12.1) and Dressel 20 amphorae (WCM fabric 42.1).

The pottery from the cultivation layer (Phase 2) was generally contemporary with the pottery from Phase 1. The samian ware from the cultivation layer in Area 3 (3001), for example, dated to the later 2nd century. One exception was a sherd from a BB1 cooking pot decorated with obtuse cross hatch burnish from Area 1 (2003). This provided the only evidence of late 3rd- to 4th-century activity. From the ceramic evidence, it is possible that the cultivation layer could have started accumulating any time from the late 3rd century onwards. There certainly appears to have been a change in the depositional processes on the site by the late 3rd to 4th century. The pottery from the Phase 2 cultivation layer was much more fragmentary than the Phase 1 pottery (Table 2.2, average sherd weight). This, and the fact that that most of it appears to be residual, indicates that the pottery from the cultivation layer is all redeposited. It is not possible to say whether it derives from subsequent disturbance of Phase 1 contexts or was brought onto the site from elsewhere. No joins were noted between the phase assemblages, but this may not be significant given the level of fragmentation in Phase 2.

Histograms have been produced to analyse the composition of the assemblage by fabric, quantified by both weight (Fig. 2.4) and rim EVE (Fig. 2.5). Perhaps the most marked feature of these histograms is the biased proportion of amphorae represented when the fabrics are quantified by weight. Most of the amphorae (17 sherds, 2401g) came

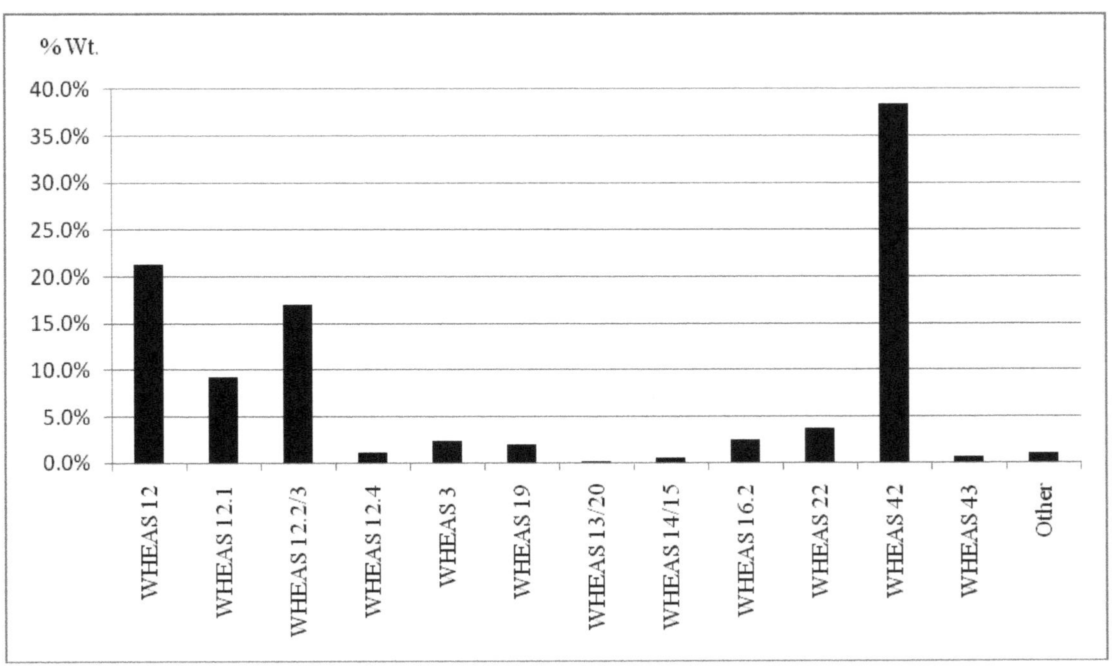

FIGURE 2.4 ROMAN POTTERY ASSEMBLAGE PROFILE BY FABRIC (% WEIGHT)

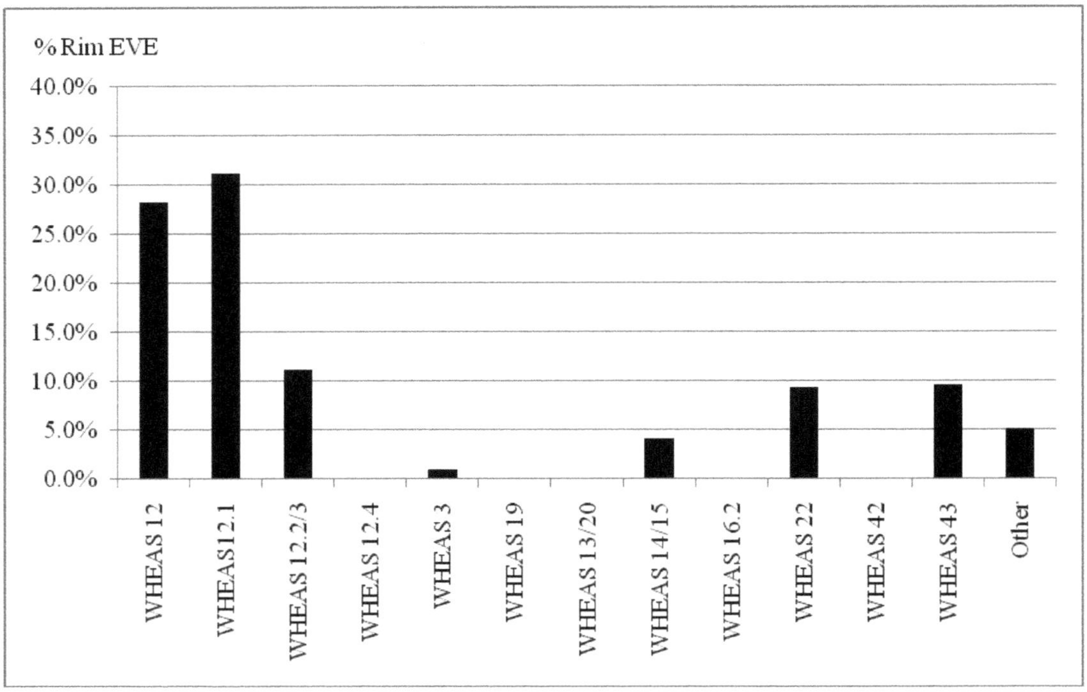

FIGURE 2.5 ROMAN POTTERY ASSEMBLAGE PROFILE BY FABRIC (% RIM EVE)

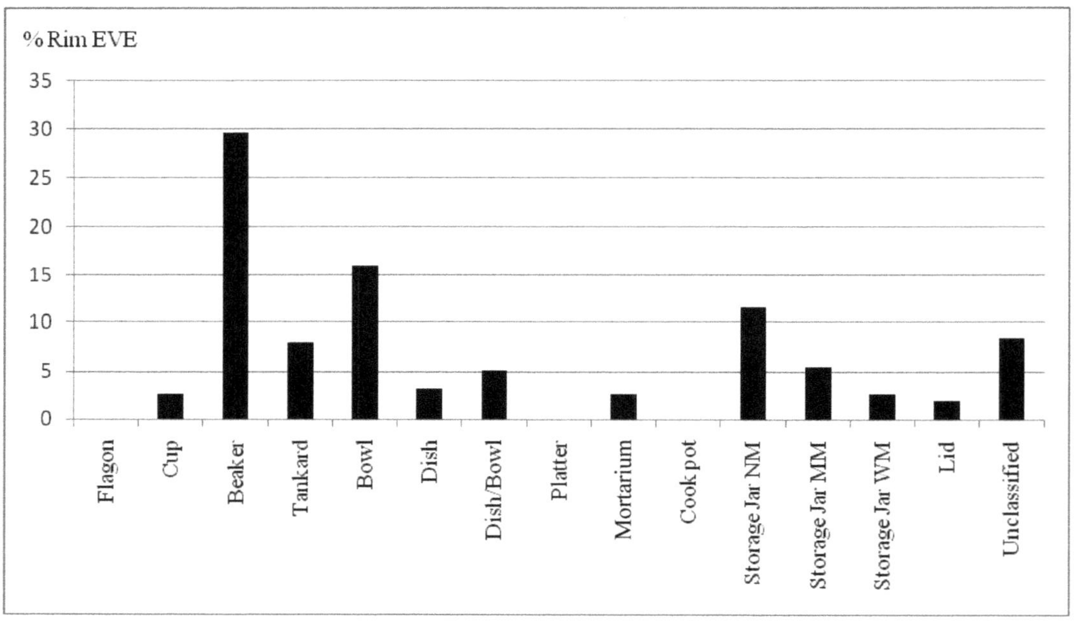

FIGURE 2.6 ROMAN POTTERY ASSEMBLAGE PROFILE BY VESSEL CLASS (% RIM EVE)

from a single pit (F10). The associated pottery included fragments from an organic-tempered Severn Valley ware storage jar and various beaker rims. Despite this obvious bias, the charts do give a useful indication of the range of fabrics present. The relatively high proportion of reduced Severn Valley ware is consistent with the evidence from 2nd-century contexts at Sidbury (Darlington and Evans 1992, 45). While it is difficult to draw firm conclusions from such a small sample, the profile of the assemblage by vessel class (Fig. 2.6) is perhaps worth noting. An emphasis on beakers, bowls and dishes, rather than jars, could reflect the type of activity on the site. A much narrower range of forms came from the cultivation layer than the Phase 1 assemblage. This must, in part, reflect the higher level of fragmentation, which made quantification of forms more difficult. There does, however, seem to be an emphasis on jars from the cultivation layer. Jars represent 24% by rim EVE in the Phase 1 assemblage, but 49% by rim EVE in Phase 2. This may reflect functional trends on the site, but as the Phase 2 assemblage is mainly residual and its derivation uncertain it is impossible to draw any firm conclusions.

SAMIAN WARE *Steven Willis*

A total of twelve sherds of samian pottery (*terra sigillata*) was submitted for identification and dating. The catalogue presented below adheres to a consistent format. Sherds are listed by context. The following data are given: sherd count, sherd type (eg, rim or body; there are no base/ footring sherds), source (South Gaulish is abbreviated to SG, Central Gaulish to CG), vessel form (where identifiable), weight of sherds in grams, the percentage of any extant rim (ie, the RE figure, where 1.00 would represent a complete circumference) and the rim diameters, and an estimate of the date of the item in terms of calendar years (this being the date range of deposits with which like pieces are normally associated). The presence of other features such as burning and sherd joins is noted.

Catalogue

F3, context 2002

1 Body, SG La Graufesenque, from a decorated bowl, either Drag. 29 or 37, 1g, *c* AD 60–100. A blurred bead border is discernible.
2 Body, CG Lezoux, Drag. 18/ 31R, 2g, *c* AD 120–50.
3 Rim, CG Lezoux, Drag. 18/ 31, 6g, RE: 0.08, Diam. 180mm, *c* AD 120–50.

Context 3001

4 Rim, CG Lezoux, probably Drag. 33, 1g, RE: 0.03, Diam. *c* 130mm, *c* AD 140–200.
5 Body, CG Lezoux, from a bowl or dish, 4g, *c* AD 140–200.
6 Rim, CG Lezoux, Drag. 31, 8g, RE: 0.08, Diam. 170mm, *c* AD 150–200.

Context 3002

7–8 One rim sherd and a conjoining body sherd, CG Lezoux, Drag. 18/ 31, 4g, RE: 0.05, Diam. 160mm, *c* AD 120–50. Both burnt.

F11, context 3006

9 Body, CG Les Martres-de-Veyre, Drag. 18/ 31, 4g, *c* AD 100–20/ 30.
10 Rim, CG Lezoux, Drag. 33, 7g, RE: 0.08, Diam. 130mm, *c* AD 120–40.
11 Rim, CG Lezoux, Drag. 18/ 31, 7g, RE: 0.01, Diam. 180mm, *c* AD 120–40.
12 Body, CG Lezoux, form not identifiable, 0.5g, *c* AD 120–200.

Discussion

This assemblage of samian ware is too small to permit any detailed discussion, and in any case comes from a range of contexts within which much of it occurs as residual material. Nonetheless, the balance of the assemblage in terms of sources and chronology is probably representative for this specific area of Worcester. The predominance of 2nd-century Lezoux sherds, together with a presence of some earlier samian, is unsurprising. The chronology of the samian sits comfortably with the dating of the coarse pottery from the four contexts within which the samian is present. In particular, the samian evidence from F3 context 2002 is consistent with the suggested dating of that context, on the basis of other indicators, as assignable to the first half of the 2nd century. All the sherds are small, with the average sherd weight falling below 4g and this may, in part, reflect the residual status of a proportion of the group. On the other hand, the sherds are comparatively unabraded and show very little sign of weathering. The soil environments of this part of Worcester seem, therefore, conducive to the survival of samian vessel surfaces and decoration, which are key elements for both dating this pottery and, in turn, early and mid Roman contexts.

CONCLUSION

This small assemblage of Roman pottery adds to the existing body of data recorded from Worcester, for example, the major excavations at Sidbury (Darlington and Evans 1992), the Magistrates's Court (Jones and Vyce 2000) and, more recently, Deansway (Dalwood and Edwards 2004). The quantity of 1st-century material is paralleled elsewhere in Worcester, with significant levels having been noted in particular on the Deansway site. As more work is done in Worcester and more quantified data are published, the character and distribution of these early assemblages need to be explored. Future work will be enhanced by the recent and imminent publication of a number of early Roman sites that may be pertinent to understanding early Roman Worcester in its wider regional context. These include *Ariconium,* Herefordshire (Willis 2000); Alcester (*Alauna*), Warwickshire (Lee *et al* 1994; Booth and Evans 2001); Metchley, Birmingham (Greene *et al* 2001; Hancocks 2004; Willis 2004); and Wroxeter (*Viriconium*), Shropshire (Timby *et al* 2000; Darling 2002).

MEDIEVAL AND POST-MEDIEVAL POTTERY *C Jane Evans, with specialist identification by Stephanie Rátkai*

The excavations also produced seven sherds of medieval pottery and 38 sherds of post-medieval pottery. The medieval pottery included Worcester-type unglazed ware (Hurst and Rees 1992, fabric 55), Worcester-type sandy glazed ware (*ibid,* fabric 64.1) and Malvernian unglazed ware (*ibid,* fabric 56), all dating to the 12th or 13th century, together with sherds of glazed Malvernian ware (*ibid,* fabric 69) dating from the 14th to 16th century. The post-medieval pottery included typically 17th- and 18th-century coarse wares, including slip wares and black-glazed wares, as well as characteristically 19th-century wares. Of potential interest were fragments from a saggar used in porcelain production (context 1001) and a fragment of unglazed porcelain (context 1002). Waste from Royal

Worcester porcelain production is fairly ubiquitous on sites in Worcestershire, sometimes being used as hardcore. However, the St Martin's Gate site is very close to one of the smaller porcelain works in Worcester, owned by the Grainger family (Pat Hughes pers. comm.), which could make these finds more significant.

ENVIRONMENTAL EVIDENCE

Charred plant remains *Marina Ciaraldi*

Introduction

A sampling programme was designed for the recovery of biological remains from the excavated features. Soil samples of 201 were taken from all datable features, in accordance with the guidelines outlined in the 'On-site Guide to Environmental Sampling and Processing' (BUFAU, Procedure no. 2). A monolith for the recovery of pollen was also taken from the southeast-facing section of Area 3 (see Fig. 2.3 and Greig, this chapter).

Methodology

Six soil samples were processed by using bucket flotation. The light fraction (flot) of the soil was recovered using a 500µm sieve while the heavy fraction (residue) was recovered on a 1mm mesh. The residues were sorted by eye, and the flots were scanned under a low-power stereomicroscope. The identification of the seeds was carried out by comparison with modern reference material. The taxonomic nomenclature for the wild species follows Stace (1991), while that of the cereals follows Zohary and Hopf (1994).

Identification

The plant macroremains recovered from the samples, though generally well preserved, presented some distortion due to their exposure to high temperatures (especially in the case of cereal grains).

Grains of spelt (*Triticum spelta* L.) and bread wheat (*Triticum aestivum* s.l.) are notoriously difficult to distinguish (Hillman *et al* 1996). In the case of the plant assemblage from St Martin's Gate, the grains of spelt had a clear, more general 'slim' aspect, with straight sides and a low lateral profile (Jacomet 1987). In a few cases it was possible to identify the wheat grains as bread wheat (*Triticum aestivum* s.l.) on the basis of their overall rounded shape and by the lack of a dorsal ridge. Grains with a short and rounded appearance and a rather steep embryo scar were identified as club/ bread wheat (*Triticum aestivum-compactum* Schiem.).

Forklets and glume basis were damaged, with only the basal part of the glumes preserved, and for the most part they could only be identified as spelt/ emmer (*Triticum dicoccum/ spelta*), as none of the features characteristic of the two species was preserved. In a few cases, however, a highly pronounced primary keel and secondary veins were visible, and these could be assigned to spelt (Jacomet 1987).

The presence of twisted grains of barley suggested that six-row barley (*Hordeum vulgare* L.) was present in the assemblage. Many detached coleoptiles were also identified, possibly belonging to cereals, on the basis of their size.

Discussion

The identifications are listed in Table 2.5. The samples discussed are from features which are all within Phase 1, dating broadly to the 1st to 3rd century. Two of them were collected from pits (F10/ 3004 and F13/ 3008), while one was from a potential ditch fill (F6/ 2007).

All the samples examined contained a mixture of cereal grains and crop processing by-products, while the presence of fruit plants, such as *Prunus* sp. or hazelnut (*Corylus avellana* L.), was only occasional.

The cereals identified include species commonly present on Roman sites, with spelt (mainly represented by chaff) largely predominant. A few grains of free-threshing wheat and bread/ club wheat (respectively *Triticum aestivum* s.l. and *Triticum aesitvum/ compactum*) were also identified. Often free-threshing wheat is sparsely represented in charred plant assemblages due to the different sequence of crop processing stages and a low chance of contact with fire. It is probable that the hexaploid rachis internodes identified in the assemblage belong to bread wheat, as these tend to be the only by-product of free-threshing wheat. The identification of emmer (*Triticum dicoccum* Schübl.) is based on the presence of emmer-like characters on a few forklets and glume basis and has to be considered uncertain. The presence of germinated grains of wheat and barley and numerous detached coleoptiles is also noticeable. These are often interpreted either as evidence of the production of beer or as an indication of bad storage conditions. In the case of the present assemblage, both interpretations are possible.

The group of wild plants includes species of weeds often present in plant assemblages from Roman sites in this region, such as Deansway and Blackfriars (see Moffett 1987, 1995, 2004b) or Glebe Farm and Causeway Lane, Leicester (Monckton 1999b and c). The group includes a mixture of by-products characteristic of fine sieving and final stages of cleaning, such as large seeded grasses like *Lolium* sp. or brome grass. The final stages of cleaning take place either just before consumption, or before storage. High numbers of large seeded grasses have already been reported from Roman deposits in Worcester, for instance at Deansway (Moffett 1995, 2004b) and Blackfriars (Moffett 1987). The similarity in species composition and numerical proportion suggests that similar agricultural practices were adopted on sites in this area of the West Midlands. The proximity of the Worcester sites could also suggest

Common name	Feature/ Context		F10/ 3004	F6/ 2007	F13/ 3008
	Area		3	1	3
	Vol. soil processed		13	15	13
	Seeds/ litre		33	0.9	3.6
	Cereals				
spelt	*Triticum* cf. *spelta*	g	4		
bread wheat	*Triticum aestivum* s.l.	g			4
bread/ club wheat	*Triticum* cf. *aestivum/ compactum* germinated	g	1		
wheat	*Triticum* sp.	g	4		2
wheat	*Triticum* sp. Germinated	g	4		
barley	*Hordeum vulgare*	g	2		2
barley	*Hordeum vulgare* germinated	g	1		
six-row barley	*Hordeum vulgare* twisted	g	1		
barley	*Hordeum vulgare* hulled	g	1		
barley	*Hordeum vulgare* tail	g	1		2
cereals	Cereals	g	22	5	8
cereals	Cereals – germinated	g	2		
cereals	Cereals – pitted	g	2		
cereals	Cereal/ large *Poaceae*	col	18		
	Chaff				
emmer/ spelt	*Triticum dicoccum/ spelta*	glb	35		5
emmer/ spelt	*Triticum dicoccum/ spelta*	f			3
spelt	*Triticum spelta*	glb	68		5
spelt	*Triticum spelta*	gl	3		
spelt	*Triticum spelta*	f	1		
hexaploid wheat	*Triticum* sp. Hexaploid	ri	5		
wheat	*Triticum* sp.	base fork	20		
wheat	*Triticum* sp.	ri	23		
barley	*Hordeum vulgare*	ri	1		
	Weeds				
fat hen goosefoot	*Chenopodium album*	s	2		
grass vetchling	*Lathyrus nissolia*	s	1		
vetch/ tare	*Vicia/ Lathyrus*	s	15	3	3
small leguminosae	*Medicago/ Melilotus/ Trifolium*	s	3		1
knotgrass	*Polygonum aviculare*	s	1		
knotgrasses	*Polygonum* sp.	s			1
sheep's sorrel	*Rumex acetosella*	s	3		1
docks	*Rumex* sp.	s	8		
cleavers	***Galium aparine***	s	3		
bedstraws	*Galium* sp.	s	2		
daisy family	*Asteraceae* (*Alchemilla* type)	s	1		
spike- rushes	***Eleocharis uniglumis/ palustris***	s	3		
club-rush	*Scirpus* sp.	s	1		
sedges	*Carex* sp.	s	1		
oats	*Avena* sp.	a	2		
rye-grasses	*Lolium* sp.	s	2		
meadow-grasses	*Poa* sp.	s			1
soft/ rye brome	*Bromus hordeaceus/ secalinus*	s	102		
thymothy grass	*Phleum pratensis*	s	12		
heath-grass	*Danthonia decumbens*	s			1
grasses	*Poaceae*	s		3	
grasses	*Poaceae Poa* type	s	42		1
small grasses	small *Poaceae*	ci/ cn	2		
large grasses	large *Poaceae Lolium* type	cn			6
	Other				
hazelnut	*Corylus avellana*	sh	2	1	
	Prunus sp	st		1	1
	Indet	s	2		

TABLE 2.5 PLANT REMAINS FROM ST MARTIN'S GATE

that the cereals were cultivated under similar ecological conditions, if not in the same field. The presence of species such as *Carex* sp., *Scirpus* sp. and spikerush (*Eleocharis uniglumis/ palustris*) in other Roman assemblages from Worcester (Moffett 1987 and 1995) would support this idea. These plants are not typical cornfield weeds but they are often found in the vicinity of drainage channels or poorly-drained areas of fields. The common finding of spikerush seeds amongst charred crop waste suggests that it was a weed associated with cereals in the past (Jones 1988a).

Heath grass (*Danthonia decumbens* L.), a plant nowadays found in heaths and acid grasslands, is thought to have been a common weed of ard-cultivated fields (Hillman 1981 and 1982b). This species seems to have been particularly common in Iron Age/ Roman plant assemblages from Wales (Hillman 1982a) and northeast England (van der Veen 1992). Hillman (1981) suggested that heath grass could be an indicator of the use of ards, rather than mouldboards, for ploughing. Ards do not cut the soil as deeply as mouldboards, allowing the permanence of perennial plants in cultivated fields. Statistical analysis of plant assemblages from the northeast of England has shown that heath grass is always associated with a spelt crop (van der Veen 1992). This might have been the case for the Midlands, but more evidence is needed to support this.

The plant assemblage from St Martin's Gate is interesting not only for the information that it provides on Roman agricultural practices in Worcester, but also in relation to other aspects of the archaeological evidence. Quantities of metal slag, possibly deriving from iron smelting, were recovered from several of the Phase 1 features. Slag and spheroidal hammerscale were also present in all the samples examined. In a sample from a pit in Area 3 (F10/ 3004), the evidence of burning *in situ* and the presence of charcoal with fragments of slag attached (see Gale, this chapter), seems to suggest a direct relation between the charred plant remains and the activity of smelting. Barker (1968–69) reported on the presence of large heaps of slag as evidence of a Roman industry of iron smelting in Worcester. Later, Carver (1980, 26) suggested that the presence of a nearby river would have been ideal for any industrial establishment in this area of the town.

The evidence suggests that the charred plant assemblage could possibly be interpreted as relating to metalworking activities, perhaps as fuel. Evidence of the use of crop by-products as fuel or as tinder is already known from other Roman sites in the Midlands. The presence of large assemblages of spelt chaff associated with corn driers, for instance, is well known (Bowker 1982 and 1983; Nye and Jones 1994; Jackson *et al* 1996; Letts 1996; van der Veen 1989; Monckton 1999a; Moffett and Ciaraldi 1999, 170). In particular, evidence of the use of fine sieving by-products as tinder comes from Catsgore (Hillman 1982b), Tiddington (Moffett 1986) and Deansway (Moffett 1995). At Stowmarket (Murphy 1989) and at Postwick (Murphy 1997) the charred remains of fine sieving waste were associated with pottery kilns. Similar evidence of the use of crop processing by-products as fuel during Roman times is also well attested in temperate and arid regions (van der Veen 1999).

The only case known to the author of the use of cereals in iron-smelting activities comes from Denmark, from a Late Roman-Germanic Iron Age site (Mikkelsen 1997), although in this case the entire crop was used as fuel. Excavations of iron-smelting furnaces in Denmark have revealed that twigs and branches, as well as straw, were used to retain the charcoal and the bog ore in the furnace while the initial iron extraction was carried out (Mikkelsen 2003). It is possible that by-products from crop processing were used as fuel, either for a particular part of the ironworking process (for instance to retouch metal objects), or because there was a more general scarcity of wood fuel, as perhaps indicated by the evidence from the charcoal remains (see Gale, this chapter).

CONCLUSION

Despite the small size of the plant assemblage from St Martin's Gate, its importance is unquestionable. The species of weeds present in the assemblage are found in most of the plant assemblages from Roman sites in the region. This raises the question as to whether this can be considered to be evidence for the existence of homogenous agricultural activities in the region during Roman times. The evidence suggests that there may be other possible common practices among sites within the region, such as the use of crop waste for fuel in association with different 'industrial' activities. However, although the regular use of crop by-products as fuel or tinder in Roman Britain is now well attested, the scale at which this occurred is less clear. It is uncertain, for instance, whether the use of crop by-products is limited to producer sites or if it was part of a long-distance exchange system. It is to be hoped that future work will be able to address this problem.

POLLEN *James Greig*

SUMMARY

Pollen samples from a pit of Roman date and a cultivation layer of probable Roman date provided only a little poorly-preserved pollen, which might be indicative of an open, occupied place, by analogy with a better-preserved assemblage from excavations at Sidbury (Sawle 1977; Darlington and Evans 1992), also in this part of Worcester.

INTRODUCTION

Pollen samples were collected from the southeast-facing section of Area 3 (see Fig. 2.3 for location of sample). This sample included the organic fill of a pit (F11/ 3006) and the overlying cultivation layer (3001).

METHODOLOGY

Two monolith tins were collected, to sample the section of pit fill (F11/ 3006) and the overlying cultivation layer

(3001, Fig. 2.3). Small sub-samples for pollen analysis were taken every 50mm throughout this section. Two pollen samples from 350mm, representing the cultivation layer, and from 700mm, representing the pit fill, were processed using the standard method. Sub-samples of approximately 10mm^3 were dispersed in dilute NaOH and filtered through a 70µm mesh to remove coarser material, which was then scanned under a stereo microscope. The finer organic part of the sample was concentrated by swirl separation on a shallow dish. Fine material was removed by filtration on a 10µm mesh. The material was acetolysed to remove cellulose, stained with safranin and mounted on microscope slides in glycerol jelly. Counting was done with a Leitz Dialux microscope. Identification was carried out using the author's pollen reference collection, seen with a Leitz Lablux microscope. Standard reference works were used, notably Fægri and Iversen (1989), and Andrew (1984).

RESULTS

The pollen types have been listed in Table 2.6, using the nomenclature and order of the taxa according to Bennett (1994) and Kent (1992) respectively. Pollen was poorly preserved and sparse, so outline counts were carried out based on five traverses of each pollen slide, and a scan of the rest of the slide under low power. Most of the pollen was of *Lactuceae*, a group which includes dandelions, hawkbits and sow thistles, a pollen type that is so robust that it is the last to decay away. It therefore seems that conditions were not suitable for the preservation of most pollen. *Centaurea nigra* (knapweed) was present in the sample from 700mm, another robust pollen grain. Other pollen grains such as *Poaceae* (grasses), *Cerealia* (cereals) and *Chenopodiaceae* (goosefoot, fat hen, orache) were pale and thin from partial decay. They show open grassy or weedy conditions. *Alnus* (alder) pollen was also found in the sample from 350mm.

A few grains were found in a very fresh state such as *Tilia* (lime), and are thought to be modern, as the lime trees were flowering at the time of excavation, in mid July.

Spores	0.35m	0.70m	Common name
Pteridium	1	1	bracken
Pollen			
Alnus	2	-	alder
Chenopodiaceae	+	-	goosefoot
Tilia	+	-	lime (probably modern)
Centaurea nigra	-	2	knapweed
Lactuceae	9	13	a group of composites
Anthemis-tp.	-	2	mayweeds etc.
Poaceae	2	-	grasses
Cerealia-tp.	1	-	cereals

TABLE 2.6 POLLEN AND SPORES FROM F11

CORRELATION WITH OTHER SITES

A useful comparison to St Martin's Gate is the site at 23–29 Sidbury, Worcester, which was excavated by John Sawle in 1976–77 (Sawle 1977). A large ditch, possibly dating from the 1st century, contained plenty of good pollen, with around 20 pollen taxa, showing very open, occupied surroundings with grassland and weeds, and a trace of cereals (Greig, unpublished data; Greig 1982). It is possible that the large pit (F11) at St Martin's Gate originally had a similar pollen assemblage, but only the traces have survived. The good pollen preservation at Sidbury demonstrates the value of looking for pollen at sites such as this, as useful results can be obtained. However, due to the poor preservation from the samples assessed at St Martin's Gate, it was not considered to be worthwhile to carry out any further analysis, as this would have been unlikely to yield any more information.

CHARCOAL *Rowena Gale*

INTRODUCTION

Charcoal from two pits (F10 and F13) and a ditch (F6) was analysed. The presence of abundant metalworking slag suggests that the charcoal may represent associated fuel deposits.

MATERIALS AND METHODS

Bulk soil samples were processed by flotation and sieving using 1mm and 500 micron meshes. The resulting flots and residues were scanned under low magnification and the charcoal separated from plant macrofossils. Charcoal fragments measuring >2mm in cross-section were considered for species identification.

The charcoal was rather poorly preserved and some fragments were vitrified. The charcoal was prepared for examination using standard techniques: fresh transverse surfaces (TS) were exposed by fracturing and scanned using a x20 hand lens. The charcoal was then sorted into groups based on the anatomical features observed on the TS. Representative fragments from each sample were selected for detailed study at high magnification. Transverse (TS), tangential (TLS) and radial planes (RLS) were prepared and the three surfaces supported in washed sand. The wood structure was examined using a Nikon Labophot microscope at magnifications of up to x400 and matched to prepared reference slides of modern woods. Where possible, the maturity (ie, heartwood/ sapwood) of the wood was assessed and the number of growth rings was recorded.

RESULTS

The results of the charcoal analysis are summarized in Table 2.7 and discussed in the following text. In the table, heartwood is indicated as (h), sapwood as (s), and twiggy fragments as (t). The anatomical structure of the

Sample	Context	*Acer*	*Betula*	*Corylus*	*Fraxinus*	*Pomoideae*	*Prunus*	*Quercus*	*Salicaceae*	*Ulmus*
Area 1, linear feature F6										
2	2007	-	-	1t	3	1	-	21s, 9h	-	1s
Area 3, pit F10										
4	3004	21	-	-	-	53	55	16h	-	-
Area 3, pit F13										
6	3008	2	1	2	2	2	-	17s, 26h	1	1

Key h = heartwood; s = sapwood; t = twig

TABLE 2.7 CHARCOAL FROM ROMAN IRON-MAKING CONTEXTS

charcoal was consistent with the taxa or groups of taxa given below. It should be noted that some related taxa are anatomically similar and secure identification to genus level may be difficult or impossible. These include members of the Pomoideae (*Crataegus, Malus, Pyrus* and *Sorbus*) and Salicaceae (*Salix* and *Populus*). Where a genus is represented by a single species in the British flora this is named as the most likely origin of the wood, given the provenance and period, but it should be noted that it is rarely possible to name individual species from wood features, and some exotic species of trees and shrubs were introduced to Britain from an early period (Godwin 1956; Mitchell 1974). Classification follows that of *Flora Europaea* (Tutin *et al* 1964–80).

Taxa identified

Aceraceae — *Acer campestre* L., field maple
Betulaceae — *Betula* spp., birch
Corylaceae — *Corylus avellana* L., hazel
Fagaceae — *Quercus* spp., oak
Oleaceae — *Fraxinus excelsior* L., ash
Rosaceae — Subfamilies:
Pomoideae, which includes *Crataegus* spp., hawthorn; *Malus* sp., apple; *Pyrus* sp., pear; *Sorbus* spp., rowan, service tree and whitebeam. More than one species may be represented in the charcoal
Prunoideae *Prunus spinosa* L., blackthorn
Salicaceae — *Salix* spp., willow, and/ or *Populus* spp., poplar
Ulmaceae — *Ulmus* spp., elm.

Area 1

The potential linear feature (F6) was possibly the construction trench of a wall. Subsequent deposits in its fill (2007) included large quantities of metalworking slag and pottery. Charred seeds, cereal grains and charcoal were sparse. The charcoal was composed mainly of fragments too small to identify. Larger pieces included oak (*Quercus* sp.) heartwood and sapwood (probably from fairly juvenile wood), ash (*Fraxinus excelsior*), elm (*Ulmus* sp.) sapwood, a member of the hawthorn/ *Sorbus* group (Pomoideae) and a twiggy piece of hazel (*Corylus avellana*), charred diameter 3mm. A small piece of coal-like material was also present.

Area 3

Charcoal associated with metalworking waste was also examined from two pits (F13 and F10). The sample from the fill of one of the pits (F13/ 3008) consisted of fragments up to 10mm in the longest axis. Despite the poor condition of the charcoal, a wide range of taxa was identified, including oak (*Quercus* sp.) sap and heartwood, maple (*Acer campestre*), ash (*Fraxinus excelsior*), hazel (*Corylus avellana*), birch (*Betula* sp.), elm (*Ulmus* sp.), willow/ poplar (*Salix* sp./ *Populus* sp.), and members of the hawthorn/ *Sorbus* group (Pomoideae). Many pieces of oak had undergone vitrification, a glass-like condition brought about through exposure to temperatures exceeding 800o C (Juliette Prior pers. comm.), during which cell walls become plastic and fuse together with consequent loss of recognisable structures. Some small fragments rather similar in appearance to the vitrified charcoal probably derived from coal. Charred cereal grains and chaff and weed seeds were also present.

The other pit from which material was analysed (F10) showed signs of *in situ* burning. Charcoal was comparatively abundant, with fragments of up to 20mm in the longest axis, and charred cereal grains, chaff, and weed seeds were also frequent. The larger fragments of charcoal mostly originated from blackthorn (*Prunus spinosa*) (one piece of blackthorn stem carried a thorn) and members of the hawthorn/ *Sorbus* group (Pomoideae). Oak (*Quercus* sp.) and maple (*Acer campestre*) were also present. Roundwood diameters of blackthorn were estimated (from fragmented radial sections) to have been in excess of 50mm when living. The pieces examined included 16 growth rings, which varied in width but which did not suggest coppiced wood. Some fragments of blackthorn (*Prunus spinosa*) charcoal and bark (unidentified but morphologically similar to blackthorn) were attached to metal slag.

DISCUSSION

Analysis of the charcoal supports the suggestion that the area was a focus of industrial activities in the Roman period, particularly metal-smelting. Charcoal from the fills of a ditch (F6) in Area 1 and two pits (F10 and F13) in Area 3, was closely associated with metalworking slag and thus attributed to deposited fuel residues, while vitrified charcoal in one of the pits (F13) attests to burning at high temperatures. Other materials from these features included pottery, charred weed seeds, and cereal grains and chaff from cereal processing waste. While the latter could be interpreted as dumped domestic waste, it could also represent the remains from industrial use, perhaps for tinder or fuel (see Ciaraldi, this chapter). By implication, similar domestic origins for the charcoal, although unlikely, cannot be ruled out.

The taxa identified from the fuel residues included oak (*Quercus* sp.), blackthorn (*Prunus spinosa*), members of the hawthorn/ *Sorbus* group (Pomoideae), maple (*Acer campestre*), elm (*Ulmus* sp.), birch (*Betula* sp.), hazel (*Corylus avellana*), ash (*Fraxinus excelsior*) and willow/ poplar (*Salix* sp./ *Populus* sp.). Assuming an industrial origin for these residues, it is clear that fuel (probably charcoal) was provided from a wide range of taxa. The relatively small quantity of identifiable charcoal in these samples, however, did not allow an assessment of species dominance or selection. It is interesting to note, though, the apparent importance of non-oak species, particularly in one of the pits (F10), where blackthorn and members of the hawthorn group were abundant. Blackthorn wood was not recorded from the other pit (F13) or the ditch (F6), but in this pit blackthorn charcoal was attached to metal slag, verifying its use in the smelting process.

In many areas of Britain charcoal residues from Roman metalworking sites have demonstrated a strong preference for oak or oak and hazel fuel, eg, at Bardown, Sussex (Cleere and Crossley 1995), Blakeney and *Ariconium*, Gloucestershire (Gale 2000a, 2000b), Woolaston, Gloucestershire (Figueiral 1992), Pomeroy, Devon (Gale 1999), Welwyn Hall, Hertfordshire (Gale, 1997), Scole, Norfolk (Gale, in Cowgill and McDonnell, forthcoming), Pakenham, Suffolk (Gale, in Plouviez, nd), Lefevre Walk and Parnell Road, Bow, London (Rackham *et al* 1998).

Environmental conditions and the local availability of wood for charcoal making would have been critical factors in determining the species used; locally produced charcoal was preferable since imported charcoal tended to fracture when carted over long distances (Galloway *et al* 1996). Pollen residues were scarcely present in the Roman levels and apart from alder (*Alnus glutinosa*), no other tree species were recorded (Greig, this chapter). Palynological evidence from other sites in this part of Worcester suggests that in the 1st century A.D. the district was mainly open grassland (Greig, this chapter). The wide range of species represented in the fuel residues, particularly shrubby species, such as blackthorn (*Prunus spinosa*) and the hawthorn group (Pomoideae), may reflect this woodland depletion, and evidence of cereal processing waste and possibly coal could be interpreted as a means of boosting fuel supplies.

Apart from blackthorn (*Prunus spinosa*) roundwood (see above), the charcoal was generally too comminuted to assess for evidence of the use of coppiced rods. Growth ring distribution in the fairly chunky stems of blackthorn, which probably measured about 50mm in diameter when living, was not indicative of coppice growth. It is worth noting the apparently sparse use of hazel (*Corylus avellana*), which is usually common and frequently coppiced in many British plant communities (Rackham *et al* 1998), and the non-use of alder (*Alnus glutinosa*), which was present in the pollen record (Greig, this chapter). Could this imply alternative uses for these species? The output of iron from this region is unknown but if production was concentrated in this area (see above) and relatively continuous, the rate of fuel use may have out-paced supply, thereby contributing to the general paucity of woodland.

INDUSTRIAL RESIDUES

SLAG *Annette Hancocks*

A range and variety of industrial waste was recovered from the site, the major component of which was 4478g of tap slag. This would have derived from a bloomery furnace and attests to iron smelting on, or near, the site. The material is highly dense, has a lava-like flow on its upper surface, and a blue-grey colour on fracture.

Large quantities of spheroidal hammerscale were recovered from bulk samples during wet sieving. The presence of this material from bulk samples is noted in Table 2.8. Hammerscale results from the solidification of small droplets of liquid slag expelled from within the iron during hot working. This is significant because it is often found in the immediate vicinity of the smithing hearth and anvil.

DISCUSSION *Lucie Dingwall*

ROMAN PERIOD

There was no evidence for pre-Roman activity at St Martin's Gate, but the remains from the site, although limited in extent, suggest that industrial activity was located in the vicinity in the early and middle Roman periods, possibly in association with agricultural activities. Ironworking slag was recovered from most of the Phase 1 features, although the quantities are very small in comparison to other sites in Worcester. However, the limited number of Roman features should be taken into consideration when comparing quantities of artefacts.

The ironworking activity at St Martin's Gate, dating from the 1st to 3rd century, is broadly contemporary with that from Deansway, and earlier than that from Farrier Street and Broad Street. Excavations at Broad Street and Deansway have yielded ironworking hearths and primary dumps of smelting slag (Jackson 2004), whereas at the St Martin's Gate site, no ironworking structures were recorded, no crucibles or moulds were recovered, and most of the slag occurred in dumps of material mixed with domestic rubbish, such as pottery. However, a high proportion of the slag recovered is tap slag, which could

Context /Feature	Tap slag	Fuel ash slag	Vitrified hearth lining	Undiagnostic iron working slags	Blast Furnace slag	Spheroidal hammerscale	Haematite	Iron debris	Total
1005 (W/ B)	68g	-	-	-	-	-	-	-	**68g**
1006 (W/ B)	78g	-	-	-	-	-	-	-	**78g**
2002 (F3)	280g	25g	-	716g	-	-	-	-	**1021g**
2007 (F6)	1313g	-	-	1824g	-	-	-	-	**3137g**
3000 (Tr.2)	-	-	-	-	-	-	2g	-	**2g**
3001 (TP1)	10g	-	-	-	-	X	-	-	**10g**
3001 (TP2)	2005g	42g	127g	990g	199g	-	-	92g	**3455g**
3002 (Tr.2)	417g	-	-	-	-	-	-	119g	**536g**
3004 (F10)	-	-	-	-	-	X	-	-	-
3006 (F11)	90g	-	-	-	-	X	-	-	**90g**
3008 (F13)	-	-	-	-	-	X	-	-	-
4002 (F20)	217g	-	-	-	-	-	-	-	**217g**
Total	**4478g**	**67g**	**127g**	**3530g**	**199g**	-	**2g**	**211g**	**8614g**

TABLE 2.8 QUANTIFICATION OF SLAG BY TYPE (WEIGHT IN GRAMS)

be indicative of a smelting furnace in the near vicinity. This is supported by the recovery of slag with charcoal still attached, the evidence for *in situ* burning in one of the pits, and indications of high temperatures as displayed by the vitrified charcoal. The lack of structural evidence for furnaces may be due to the small size of the area excavated, or to the fact that Roman remains have only survived in discrete areas of the site. It is possible that structural evidence for ironworking could survive in areas of the site that were not disturbed, and remain to be discovered in the future.

Of particular significance is the retrieval of spheroidal hammerscale from the pits, which has a clear association with the occurrence of smithing in the near vicinity (Starley 1995). This could be either primary smithing of the bloom, or secondary smithing of objects for manufacture or repair. Smithing was often found in the immediate vicinity of the hearth and anvil, and did not require purpose-built structures (McDonnell 1995). The lack of evidence for smithing at other Roman sites in Worcester has led in the past to the characterisation of the iron industry in Worcester as a production centre for iron that was used elsewhere (Jackson 2004). However, this apparent absence of smithing may in part be due to the fact that hammerscale was not looked for, or recognized, in samples from past excavations. More recent work suggests that smithing was widespread (James Dinn pers. comm.).

The palaeoenvironmental remains have produced important information about fuel usage in the area in the Roman period, and it is unusual to find plant remains in direct association with metalworking. This clearly suggests that the charcoal represents deposited fuel residues associated with metalworking, rather than dumped domestic waste. The availability of charcoal may have been the most important factor in determining the location of furnaces, as large quantities were needed and could not easily be transported over great distances (McDonnell 1995). The suggestion that crop by-products were being used as fuel due to scarcity of wood supplies is supported by the evidence from other Worcester sites that cereal chaff was used for fuel (eg, The Butts sites reported in this volume: Deansway, Moffett 2004a; Farrier Street, Dalwood *et al* 1994).

Although the quantities of ironworking residue recovered from the site at St Martin's Gate are relatively small, the character of the archaeological remains fits well with the emerging picture of the establishment of a major iron-smelting industry in Worcester by at least the 2nd century (Jones and Vyce 2000; Dalwood and Edwards 2004).

There was very little evidence generally for structures of Roman date on the St Martin's site, with the exception of an isolated posthole and a possible robbed wall in Phase 1. However, the recovery of domestic pottery indicates that even if people were not actually living on the site, they were living close by (animal bone is notable for its absence). The lack of structural evidence may be due to the ephemeral nature of many poorer class dwellings and industrial buildings in the Roman period (White and Baker 2000). It is also likely that the earliest buildings on the site would have been timber buildings, which are less likely to survive in the archaeological record.

It has been suggested that the Roman activity to the north of the historic city represents an industrial suburb of the small town, to the north of the postulated nucleus of the Roman settlement. However, the gathering body of evidence from various excavations and observations indicates that occupation of the town was intermittent over certain areas in the early Roman period (Dalwood *et al* 1994; Dalwood and Edwards 2004), suggesting that the

Roman settlement was diffuse in character, with a number of different foci (WCC 2007). The location of ironworking at St Martin's Gate, away from the postulated 'industrial suburb' in the north, would support this view. No evidence was found on the site of the historically documented Roman road towards Droitwich (White and Baker 2000), but the location of ironworking in the near vicinity of an important road would be consistent with the evidence found in the north of the city for such activities along the major north–south Roman road to Greensforge.

The evidence from Sidbury, Farrier Street and Deansway has indicated that these parts of the town were abandoned by the late 4th century (Darlington and Evans 1992; Dalwood *et al* 1994; Dalwood and Edwards 2004), although, unusually for Worcester, the County Education Offices in Castle Street yielded pottery dating to the late 4th to early 5th century (Dalwood *et al* 1997). The cultivation layer from Phase 2 at St Martin's Gate is paralleled elsewhere in Worcester. Such deposits, ranging in depth from 0.6–1.2m, have been identified at many other sites in the city, including Farrier Street, Deansway, the Kardonia site and Castle Street. Several of these contained medieval and post-medieval artefacts from later reworking. The presence of late 3rd- to 4th-century pottery in the cultivation layer at St Martin's Gate indicates a late 3rd century or later date for deposition. It is possible that this later pottery was redeposited from the Phase 1 activity on the site, which would extend Phase 1 into the 4th century. However, the complete absence of any material later than the 3rd century in the Phase 1 assemblage makes this interpretation less likely.

It is more probable that the cultivation layer was partially or wholly composed of material derived from elsewhere in the vicinity, and this is supported by the level of fragmentation of the pottery. It is impossible to establish this for certain, since the ceramic assemblage is too small and fragmentary to enable the identification of vessel joins between the two phases, or to be able to make conclusive statements regarding functional differences between the ceramics in the two phases, other than the higher proportion of jars in Phase 2 (see Evans, this chapter).

The presence of cultivation layers, sometimes termed 'dark earth' deposits has been taken to indicate an absence of late Roman or immediate post-Roman occupation in the area. However, it could be argued that the evidence of gardening or farming as represented by the 'dark earth' deposits must indicate some level of occupation, but the density of occupation remains a matter of debate. Any such occupation is likely to have involved the use of comparatively ephemeral buildings that would be difficult to detect in the archaeological record (Coates and White 2000, 6).

MEDIEVAL PERIOD

The lack of medieval activity is paralleled on the sites at The Police Station and Rea's Timber Yard, which also lacked any medieval features and produced very little medieval pottery; this has been attributed to the fact that these areas lay outside the limits of the planned medieval suburb (Dalwood *et al* 1994, 107). At St Martin's Gate site, the absence may be due to a combination of ephemeral structures and their subsequent destruction during the Civil War, followed by extensive ground disturbance in the later post-medieval period. Excavations at Tallow Hill/ Pheasant Street (Whitworth and Edwards 2001) demonstrated that post-medieval activity had truncated the alluvium in this area.

POST-MEDIEVAL PERIOD

The post-medieval brick walls located during the excavation and the watching brief correlate relatively closely with properties and property boundaries marked on the First Edition Ordnance Survey map of 1888. However, the position and size of the deep, north–south aligned, post-medieval linear ditch running across the site raises the possibility that this ditch relates to a Civil War earthwork, since it does not correspond to any post-medieval property boundaries marked on the historic maps. A putative Civil War ditch was identified at Farrier Street (Dalwood *et al* 1994, 108), and very tentatively interpreted as relating to one of the artillery bastions projecting from the city wall, as depicted on the plan of the 1651 defences. The upper profile of the ditch at St Martin's Gate was heavily disturbed, making it difficult to make definitive statements regarding date and original depth. However, the lower fill did not contain any pottery later than the late 17th century, and it is probable that this ditch represents earthworks relating to one of the Civil War sieges. There was no evidence for a bank, and the position and alignment do not correlate with the reinforcement of the city walls, but, as at Farrier Street, it may relate to an artillery bastion projecting from the wall. Other stretches of defensive ditch from the Civil War period have been identified at two of the other sites described in this volume (8–12 and 14–24 The Butts).

The archaeological evidence from the watching brief in the northwest corner of the site may support the documentary evidence for a tilehouse in the area in the late medieval and early post-medieval periods. A tilehouse is recorded as having stood opposite the gate in what is now Silver Street, first mentioned in 1455 and continuing in use by the Marson family into the 17th century (Hughes 1980, 285). A brick kiln is also mentioned. It is possible that the sandstone foundations recorded during the watching brief (F102 and F103) represent the reuse of earlier building foundations relating to this tilehouse, which was presumably destroyed during the Civil War, although the sandstone could have been robbed from the city walls. The dense concentration of post-medieval tiles recorded during the watching brief in this area may also be related to the tilehouse. There was no evidence for any kiln structures in the vicinity, but these may survive nearby, in areas left undisturbed during the development.

CONCLUSION

The site at St Martin's Gate, although providing limited and fragmentary structural evidence for the Roman period, yielded significant artefactual and palaeoenvironmental remains which, when studied in the context of other published work in the city, provide significant new information about Roman ironworking in Worcester and about the character of the settlement. The evidence for early ironworking is particularly important, as understanding the development of the Roman ironworking industry in Worcester has been identified as a key research objective (Dalwood *et al* 1994; Dalwood and Edwards 2004; WCC 2007). The St Martin's Gate excavation has provided firm evidence for Roman settlement in this part of the city.

The survival of probable Civil War remains is of particular interest, as few of the defensive earthworks are documented, and archaeological evidence is the only way to establish accurate locations for these features. There may also be potential for late medieval and early post-medieval archaeological remains surviving in the northwestern area of the site, undisturbed by the recent development, which could be associated with, and enhanced by, documentary evidence. The archaeological work has clearly demonstrated that despite extensive post-medieval and modern disturbance, there is significant archaeological survival in the northern and western part of the site.

ACKNOWLEDGEMENTS

The excavation was undertaken by Birmingham Archaeology (BUFAU), supervised by Laurence Jones, with the assistance of Eleanor Ramsey, Richard Cherrington and Lucie Dingwall, and the watching brief was carried out by Eleanor Ramsey and Mary Duncan. The project was managed by Lucie Dingwall for Birmingham Archaeology (BUFAU), and monitored by James Dinn on behalf of Worcester City Council. Thanks are due to Motor Design Group for commissioning the project, to Inchcape Motors for sponsoring the project, and to Craig Darrood for his assistance whilst on site. Thanks are also due to Pat Hughes for providing information for the report. The report was edited by Simon Butler and Richard Cuttler and the illustrations were prepared by Helen Moulden. Copy editing was carried out by Malcolm Hislop.

CHAPTER 3: 1 THE BUTTS

Mike Napthan

with C Jane Evans, Elizabeth Pearson, and Angie Bolton

SUMMARY

A single evaluation trench was excavated across the full depth of the plot at 1 The Butts to determine the position of the medieval city ditch and the archaeological potential of deposits between the city ditch and city wall. This evaluation established the presence of Roman and post-medieval deposits, most notably a late Roman stone-lined well. During the 2004 redevelopment this well was preserved *in situ*, and now lies in a small yard outside the footprint of the new buildings.

The fill of the well was partially excavated (to a depth of 1.5m), and proved to contain a rich assemblage of late 4th-century artefacts and building materials, which may have come from several different buildings. Predominantly this material comprised stone, brick and tile fragments, which appear to be demolition debris from a substantial Roman structure, or structures. Numerous stone flooring cubes (*tesserae*) and fragments of painted wall plaster point to a building of some pretensions. The function of the building is unknown, but the inclusion of stack bricks (*pilae*) and box-flue tiles might point towards under-floor heating for a bath house. Other noteworthy material included a stone column fragment, and a hollow box tile fragment with a shape suggestive of a voussoir from a vault or dome. Clear signs of scorching and burning suggest that the building(s) may have burnt down.

A wide range of domestic material was also present in the well fill, including fragments of two fine glassware vessels and a range of local and imported pottery vessels. Nine coins (all dated to the mid to late 4th century) were recovered, along with fragments of a copper alloy bracelet and a possible earring. Iron slag was present in abundance and presumably reflects the proximity of ironworking sites in the Roman town. Interestingly, coal was also present, and this is the first secure evidence for the use of coal in Roman Worcester. Animal bone included substantial fragments of horse, sheep/ goat and pig, and a worked tine from a very large deer antler.

The well structure itself included reused architectural elements from an earlier high quality stone building. Thus the well and its fill deposits reveal the likely presence of significant Roman buildings in the vicinity. Unfortunately, these are unlikely to have survived *in situ* to any great extent due to the cutting of the medieval city ditch, and other historical activity.

A substantial east–west aligned ditch cut the construction trench for the well. The material within the ditch contained sparse Roman material, including box tile and iron slag, all of which might have been residual. This feature may date to the late Roman or post-Roman periods, and had been truncated by the medieval city ditch.

Further Roman deposits were encountered during ground works for the new building and included a pit containing a substantial assemblage of 2nd- to 3rd-century pottery, including a near complete Severn Valley ware tankard, and fragments of an unusual polychrome snake-thread decorated glass vessel, probably imported from the Cologne area. Roman sub-soils also appeared to survive in the area immediately beneath the city wall, although they had been truncated across the rest of the site.

Medieval activity on the site was represented by the surviving original stonework of the 13th-century city wall, and by the city ditch, which was separated from the wall by a berm. This berm was up to 4m wide at the southern end of the site, and appeared to reduce in width towards the north. The post-medieval deposits primarily consisted of the post-Civil War infilling of the medieval city ditch.

INTRODUCTION

The site (NGR SO 8480 5410) lies immediately to the south of the modern roadway known as the Butts, and immediately to the north of the medieval city defences, here consisting of a length of wall and ditch (Figs 1.2, 1.3, 3.1). On the east, it is abutted by the former congregational Chapel Sunday School building (designed by Sir Aston Webb in 1888) and to the west by a 1998 tyre and exhaust centre. It is directly opposite the excavated sites at 8–12 The Butts (see Chapter 4), and just to the southeast of 14–24 The Butts (see Chapter 5).

Historically, the site may have been primarily agricultural land lying just to the north of the precinct of the Blackfriars Dominican Friary, founded in 1347. However, the 'Butts' place-name is generally associated with medieval archery practice ranges, found on the perimeters of several medieval towns and cities. By post-medieval times the area was known as the 'Gravel Butts' which may indicate that sand and gravel extraction had occurred.

Significant building development in the area only occurred after the partial slighting of the city defences in the mid 17th century. These buildings may have included a Georgian bathhouse, but its exact location is uncertain, and by the 19th century the site was part of the extensive Netherton House estate which occupied much of the Butts area (plan in WRO BA 8782/ 63 ref 899:749).

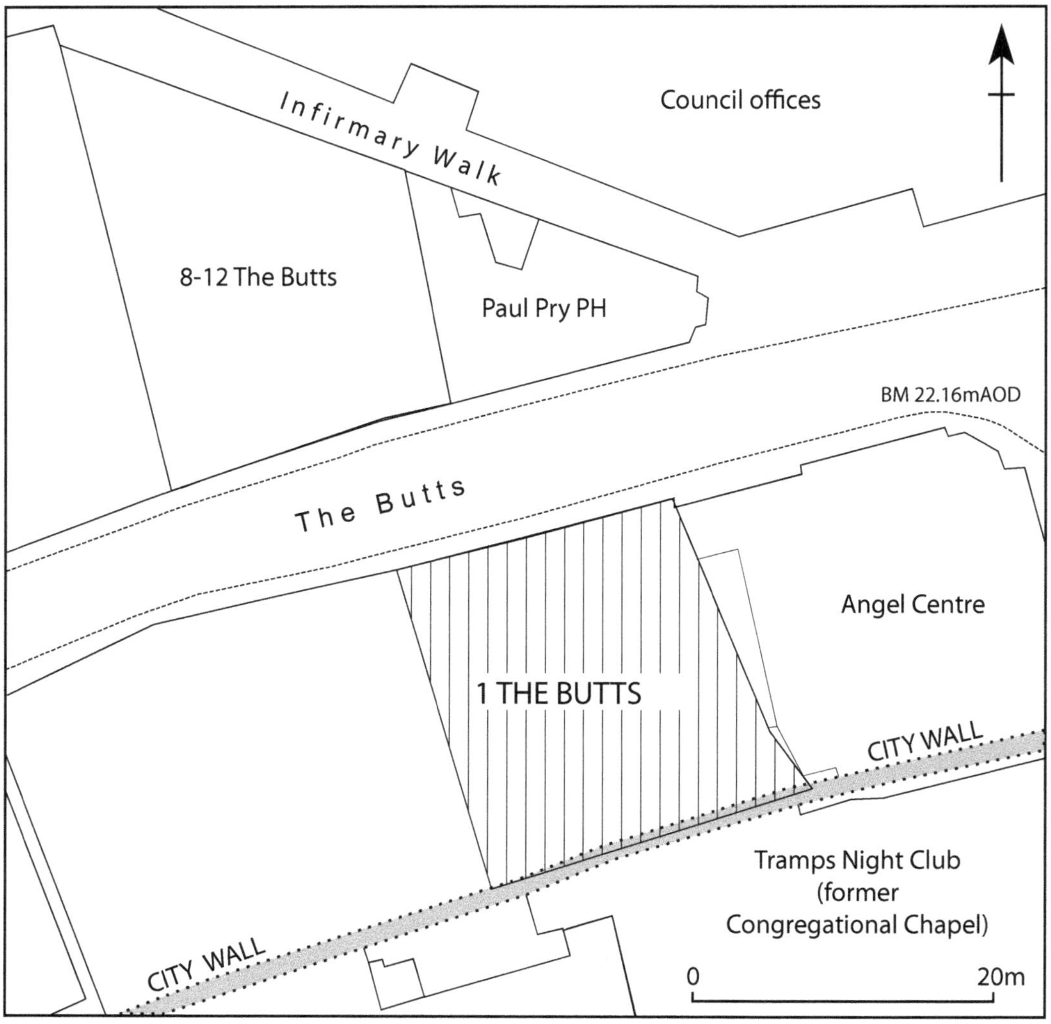

FIGURE 3.1 LOCATION PLAN

METHODOLOGY

Following archaeological building recording and demolition of the standing buildings, the site was stripped of existing hard surfacing. A single north–south evaluation trench (WCM 101108) was excavated across the berm and line of the City Ditch (Fig. 3.2). Initial stripping and excavation of 19th-century and later deposits (101) was undertaken using a mechanical excavator. At the southern end of the trench, stonework (of what subsequently proved to be a well) was observed at surface level prior to machining and this part of the trench was entirely hand excavated. The full depth of the city ditch was not excavated for practical reasons; however the approximate profile and nature of deposits was determined by hand augering. All subsequent excavation and cleaning, including the fill of the well, were undertaken by hand and the deposits recorded.

The majority of this archaeological investigation was undertaken during the evaluation stage of the project (WCM 101108) and this was limited to a single trench running from the city wall to the street frontage on the western side of the site. The exposed features were only partially excavated, the well fills being removed only to a depth of 1.5m below current ground level. Subsequent observation was limited to the monitoring of groundworks (shoring pits, piling, excavation of ground beams and service trenching). A number of small trenches (approximately 1m long, 0.5m wide and 0.4m deep) were archaeologically hand excavated where the ground beams impinged on significant Roman deposits immediately between the southern pile row and the city wall. Three principal Roman features were partially excavated – a pit (201) filled by context 202, a well (110) within a construction pit (109) and a probable ditch (106).

On completion of the evaluation, the well was backfilled with sand/ gravel, and the loose ashlar block recovered from the fill replaced on the uppermost course. The well was protected by plywood during the building works prior to the laying of paving. It is now marked on the surface by a ring of cobbles.

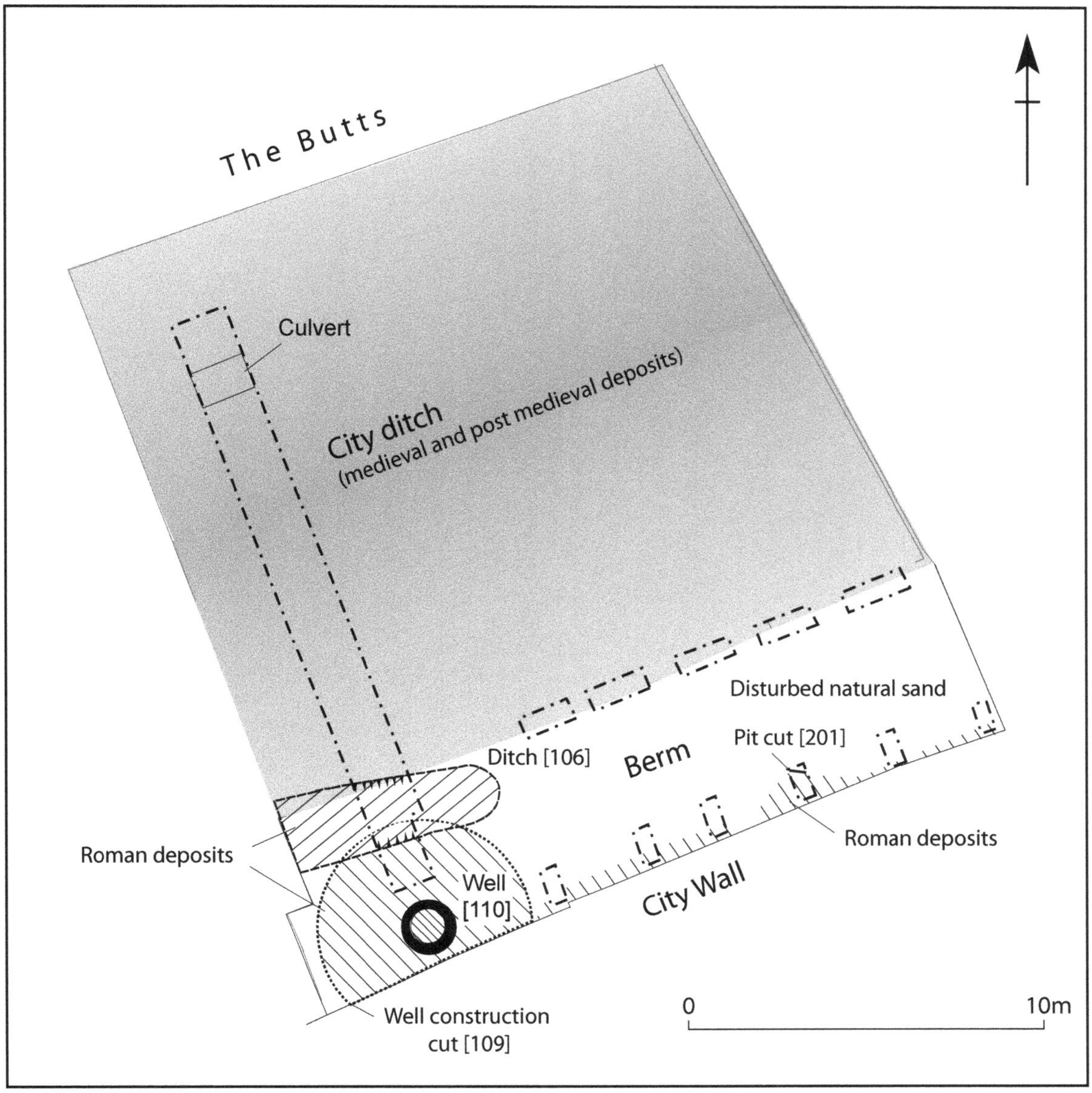

FIGURE 3.2 GENERAL SITE PLAN

RESULTS

NATURAL DEPOSITS

Fig. 3.3 shows that natural sand was present to within 0.1m of current ground levels at the southern edge of the site and 0.6m on the northern frontage, but that it had been substantially truncated in between these points by the medieval city ditch. The ditch appeared to have penetrated the sand and gravel cap completely at its deepest point, exposing the underlying marl at around 6m. The majority of observations of the natural subsoil were during piling operations by continuous flight auger and the depths of observed deposits were, therefore, very approximate within the ditch line. It was clear that the natural was rather variable and contained pockets of very wet sand over a soft sandy decayed mudstone, which only gave way to solid Keuper Marl just before the 7m limit of the piling was reached.

EARLY ROMAN DEPOSITS

The earliest observed, probably Roman, deposits consisted of an undated, light grey-brown sandy loam that survived in section beneath the city wall line. This material appeared to be a cultivated soil. It was not possible to investigate this deposit, apparently Roman subsoil, due to its location under the retaining wall, but it was clearly cut by the excavated later Roman features.

2ND AND EARLY 3RD CENTURIES

The earliest closely datable deposit examined was a cut feature (201) (Fig. 3.2) of unknown shape and dimensions filled by a dark-grey sandy loam (202) which contained a finds assemblage dating to the 2nd to 3rd century. Cut 201 extended beneath the line of the City Wall, but did not appear to extend more than 1.5m to the north. The feature was only identified in the watching brief stage

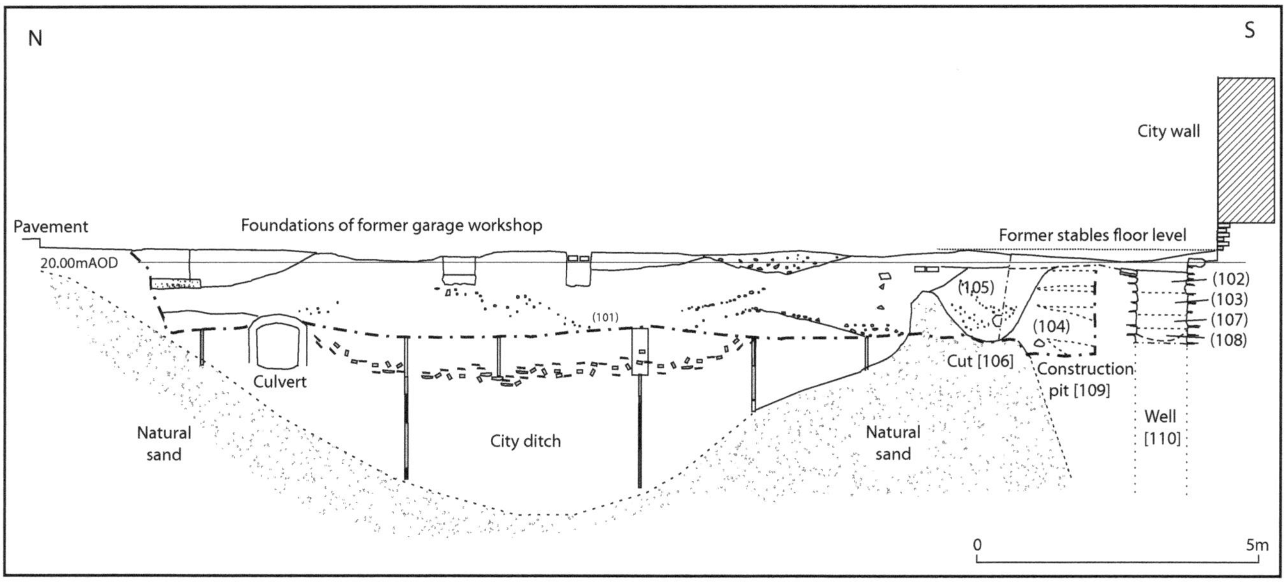

FIGURE 3.3 WEST-FACING SECTION OF EVALUATION TRENCH

of the project. Extensive disturbance during the shoring and piling operations had pushed recent rubble into the surface (as a 'piling mat') and the feature was not visible in plan. The depth of the feature appeared to be *c* 0.5m. The principal artefacts recovered from this context were a near complete Severn Valley ware tankard and a fragment of trail decorated polychrome glass, possibly from the Cologne area.

3RD TO 4TH CENTURY

Cutting the natural sand in the southwest corner of the site, between the city ditch and the city wall, was a very substantial and probably circular cut feature 109 (Figs 3.2 and 3.3) interpreted as the construction cut for the stone lined well (110). The fill of the construction cut (104) was primarily redeposited sand with tip lines of loam and clay. This material contained few datable artefacts but the small quantity of pottery recovered apparently dated the feature to the 3rd to 4th century.

LATE ROMAN OR POST-ROMAN DEPOSITS

A short length of probable ditch 106 cut the fill (104) of the construction cut for the well. The ditch fill (105) could not be closely dated, but included both iron slag and box tile fragments. Very little of the ditch fill was available for hand excavation, and the ditch appeared to terminate just to the east of the evaluation trench. It is possible that the ditch was open whilst the well was still in use, but it appears more probable that it was cut after the well fell out of use.

MEDIEVAL DEPOSITS

The city ditch was observed as a very large and deep area of silty and sandy deposits extending across the majority of the site (Fig. 3.3). The ditch was only partially excavated during the evaluation to permit recovery of its profile through hand-augering. Ditch fills at least 4.8m deep and including waterlogged deposits were identified extending to a width of more than 14m, and continuing beyond the end of the trench. The full depth of deposits could not be determined at all points due to solid obstructions (brick, stone or tile), but extended to at least 4.8m below current ground level. The upper part of the ditch fills appeared to be of post-medieval date as they contained lenses of stone and brick rubble. No dating evidence was recovered for the deeper, waterlogged, deposits, which appear to contain organic remains including timber. Whilst observation of the pile uprisings during the watching brief confirmed the presence of the ditch and the nature of the fills, the conditions of observation precluded more accurate assessment of the profile of the ditch.

THE WELL (110)

The stone lining of the well was exposed by partial excavation to a depth of ten courses of stonework (Fig. 3.3). The uppermost course (which survived up to 20.06m AOD) only partially survived. The stonework was all of fine light reddish brown sandstone ashlar, several of the blocks showing indications of re-use. The blocks had been carefully re-cut to form the curving edges of the shaft. The coursing varied slightly in depth and some blocks had been joggled to adjust the coursing of blocks of different depths and sizes. At least one block had a finely finished simple string moulding, suggesting that some of the stone had been originally used in an ornate public building or monument. The upper edges and exposed faces of the blocks had been considerably abraded by the passage of water containers, and no tooling was visible on the exposed faces. The well had apparently been truncated by post-medieval buildings, but there was no sign that the well had been otherwise disturbed since the late Roman period – should it have been exposed during construction of the city walls it is likely that the stonework would have been robbed for re-use. A single stone block (part of the structure) was recovered from the fill of the well, and this appeared to have been accidentally knocked in as the well was infilled, but it is possible that the upper part of the shaft was deliberately demolished, and that further fragments of the well structure remain in the unexcavated lower part of the well.

The material within the well (104) was excavated in four arbitrary contexts (102), (103), (107) and (108) to a depth of 1.5m (Fig. 3.3). The material of all four contexts was similar and the division into separate contexts was undertaken purely to help locate material from different depths. The remaining fills were preserved *in situ* and it was not, therefore, possible to determine the depth of the well or the date at which infilling of the shaft commenced. The abundant coin dating evidence from the excavated portion of the well points to infilling between AD 370 and 390 (see below), and the pottery appears to be consistent with this. It is important to note that the excavated deposits were clearly deposited as backfilling after the useful life of the well had ended, and were not related to its use. The diameter (between 5 and 6m) of the well construction cut (109) at the excavated level was such that it suggests that at least 5 or 6m of the well remain unexcavated.

The material from the excavated contexts will be discussed as a group, as it was clear from the consistency of the fills, the coin evidence and the presence of cross-joining sherds in different spits that the excavated portion of the fill was deposited within a very short period of time. The reason for the infilling of the well is uncertain, particularly as the lower fills remain unexcavated, but it is clear from the excavated contexts that the infilling material largely derived from the demolition and clearance of a very substantial masonry building or buildings, showing some evidence for fire damage.

FINDS

THE WELL

COINS *identified by Angie Bolton*

A total of nine coins were recovered, all from the excavated upper filling of the well (103 and 107).
One coin of House of Constantine (324–37), uncertain denomination, contemporary copy, fair condition
One coin of Magnentius (350–53) – AE 2, slightly worn
One coin of Constantinius II (337–61) AE 2, slightly worn
One coin of Constantinius II (337–61) –AE 3, fair condition
One coin possibly of Constantinius II (337–61) – AE 3, fair condition
One coin of House of Valentinian (364–78) – AE 3, poor condition
One coin of House of Valentinian (364–78) – AE 3, poor condition
One coin of unknown 4th C – AE 4, poor condition
One coin of unknown 4th C, uncertain denomination, contemporary copy, poor condition

The coins were found distributed throughout the fills and were not a single discrete deposit. The group appears to have been deposited in the 370s or 380s, with a strong probability towards a date at the end of that range as a number of the coins showed signs of significant wear. The circumstances of deposition are uncertain – the density of coins, (approximately nine per m^3 of excavated fill), suggests either accidental losses in a public area where money regularly changed hands, or a disturbed low value coin hoard, though the latter seems less likely given the range of coinage. There is very little probability that the coins were votive deposits as the well had clearly been disused and almost entirely in-filled prior to their deposition.

COPPER ALLOY OBJECTS

Three identifiable objects of copper alloy were recovered, all from the fills of well (110). The largest was a curving thin band fragment probably from a hook and loop fastening bracelet. The surviving fragment, representing about 30% of the complete object, is 58mm long, with an approximately 0.5mm thick, 1.5mm wide, rectangular section with a rounded point terminal.

A small twisted alloy wire ring approximately 8mm in diameter may have been an earring or (perhaps more likely) served a more utilitarian purpose, as Roman earrings are an uncommon find (Crummy 1983).

A fragmented and badly corroded pin, approximately 60mm in length and 1mm in diameter, appears to have been a brooch pin, though the remains of the terminal are too corroded to be certain. The remaining copper alloy fragments were mainly recovered during the sieving of soil samples and consist of several tiny flakes of decayed copper alloy sheet with no diagnostic features.

IRON OBJECTS

A considerable number of heavily corroded iron nails were also recovered from the well, representing a range of sizes up to approximately 80mm long, and would appear to have derived from non-structural timberwork, fixtures and fittings. The majority of the ironwork recovered was in the form of nails; however, the one notable iron object was a 120mm long possible hinge loop formed from a single strip of iron, bent double to form a loop at one end (Fig. 3.4). This object, from context 103, weighed 145g.

SLAG

Approximately 24kg of iron slag was recovered from the fills of the well. Roman slag is ubiquitous in this area of the city and it clearly had a high level of residuality, which means that it is not possible to determine if ironworking was still active in this area in the late 4th century on the basis of these results. Soil samples were scanned for smithing waste without definite conclusion – hammer-

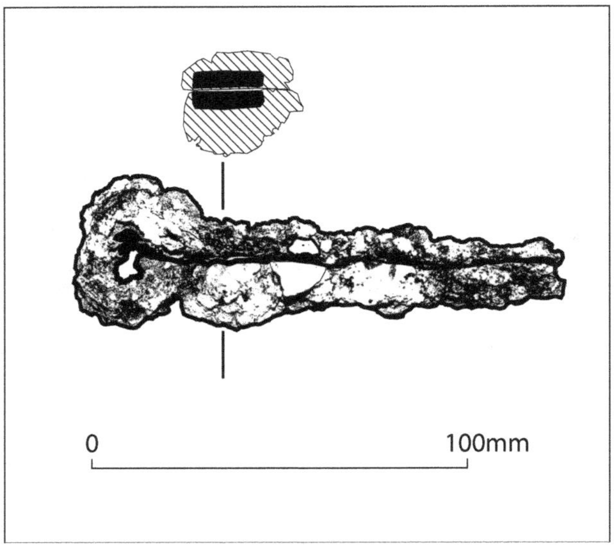

FIGURE 3.4 IRON OBJECT, POSSIBLY PART OF A HINGE, FROM CONTEXT 103

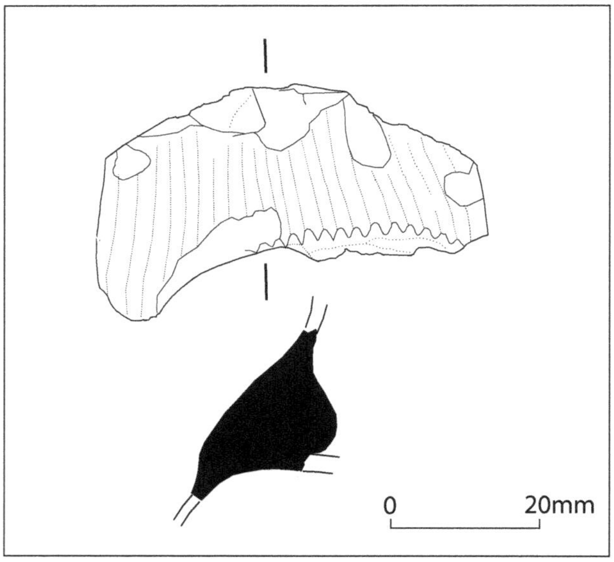

FIGURE 3.5 BLUEISH GREEN MOULDED GLASS FRAGMENT FROM CONTEXT 107

scale was present but in low concentrations, and may also have been residual.

GLASS

A small group of vessel glass was recovered from the fills of the well. Of these only the form of one vessel, a small folded beaker, was discernable. This vessel showed signs of a brown coating, possibly the remains of paint. The form would appear to be similar to that illustrated in Allen (1998, plate 19, 12), which is dated to AD 70–170. Figure 3.5 shows a moulded glass fragment from context 107, which could possibly be a ribbed bottle handle from the junction with this vessel body. It is likely to be of 1st-to early 2nd-century date, as blue green glass is predominantly 1st century or earlier (Allen 1998). It is of note that none of the identifiable glass appeared to be contemporary with the infilling of the well.

Other glass material recovered from the well included:

Context 103 – rim sherd and joining fragments of very fine possible painted colourless blown glass folded beaker – 1g.

Context 103 – Seven tiny sherds of clear glass recovered from sieving – 1g.

Context 107 – thick fragment of blueish green moulded glass, probably from junction of body and handle, fine ribbed decoration – 18g.

ROMAN POTTERY *C Jane Evans*

Introduction

The well construction cut and infill layers produced a total of 7 and 176 sherds of Roman pottery, respectively (Table 3.1). The infill layers also produced demolition debris and, amongst other finds, a number of coins. The latter provide good independent evidence for the assemblage, dating it to the end of the 4th century. This is a small but important assemblage of pottery and, at the time of writing, was one of the very few late Roman groups from Worcester, and the only sealed, well stratified and, apparently, tightly-dated late group. The latter is relatively rare in Worcester as most excavated assemblages can be expected to contain quantities of residual material. The well-preserved sherds provide evidence for the range of forms and fabrics in use in Worcester at the end of the Roman period.

Methodology

The pottery from the well (and other evaluation and watching brief contexts) was recorded using the same methodology as for 14–24 The Butts (see Evans, Chapter 5). To avoid duplication, this is summarised only within the Chapter 5 report, which also includes a list of all fabrics represented (Table 5.8).

Dating evidence

A Severn Valley ware jar dating broadly to the late 3rd to 4th century (Fig. 3.6, 1) provided a *terminus post quem* for the fill of the well construction pit (109), and thus the construction of the well. The associated pottery was less diagnostic. This included a sherd of Dressel 20 amphorae (Peacock and Williams 1986, class 25, 136–40) dating to sometime between the late 1st and 3rd century (*ibid*, 136).

The ditch cutting this feature (106) included a typically 4th-century Severn Valley ware tankard (Fig. 3.6, 2), and characteristically 4th-century fabrics such as Oxfordshire colour-coated ware and shell tempered ware. Another, near complete, 4th-century tankard was recovered during the watching brief (Fig. 3.6, 17).

The group from the well included a number of diagnostic sherds, indicating a tightly-dated assemblage. Joins were noted between the various fills, suggesting that the separate spits represent a single episode of backfilling, and, for this reason, the fills are discussed as one group. It should,

Feature	Qty	% Qty	Wt (g)	% Wt	Av. Sherd Wt (g)	% Rim	% Rim EVE
Ditch 106	9	3.2	100	1.4	11	8	1.3
Construction pit 109	7	2.5	257	3.5	37	12	2.0
Well 110	176	63.5	3600	49.2	20	459	75.1
U/ S	10	3.6	86	1.2	9	15	2.5
Evaluation	202	72.9	4043	55.2	20	494	80.9
Watching Brief	75	27.1	3278	44.8	44	117	19.1
TOTAL	**277**		**7321**		**26**	**611**	

TABLE 3.1 SUMMARY OF ALL POTTERY BY CONTEXT

however, be remembered that the well was only partially excavated; nothing is known about the earliest deposits, which may contain pottery lost in the well while it was in use. The well contained a range of other finds, including building material with evidence of burning (Napthan, this chapter) which hints at the destruction of a significant Roman building in the vicinity. The assemblage could represent pottery used in association with that structure or, as with the 4th-century assemblage from Bays Meadow villa, Droitwich, be associated with subsequent occupation (Barfield 2006, 126, 138).

The assemblage contained a number of characteristically late Roman fabrics and forms. Of particular note is the quantity of wheelmade, shell tempered ware (WHEAS Fabric 23). The dating and significance of these has been described in detail in relation to 14–24 The Butts (see Chapter 5). Only jars, with undercut rims, were found in this fabric at 1 The Butts; a diagnostically late type (Fig. 3.6, 11).

Oxfordshire products were well represented (Table 3.2), particularly the red and brown colour-coated ware. These included diagnostically late bowls and beakers (Fig. 3.6, 12–14). The latter included a number of brown colour-coated body sherds with rouletted decoration. The assemblage also contained an Oxfordshire parchment ware bowl (Fig. 3.6, 16) and an Oxfordshire white mortarium (Fig. 3.6, 15), both typically 4th-century types.

The colour-coated wares were checked for southwestern brown slipped ware, a late 4th-century ware found at Alcester and other sites in the region (Evans 1994, 147–48, fig. 67). However, no sherds matched the sample included in the County Fabric Series.

Dorset Black Burnished ware (BB1) was another common fabric (Fig. 3.6, 8–10; Table 3.2). Identifiable forms included jars and plain-rimmed dishes. The jars (Fig. 3.6, 8–9) have parallels in 4th-century or later contexts at Dorchester Greyhound Yard (Seager Smith and Davies

Common Name (National Fabric Reference code)	WHEAS Fabric Code	No.	% Qty	Wt (g)	% Wt	% Rim	% Rim EVE
Severn Valley ware, oxidised (**SVWOX 2**)	12	40	19.8	1055	26.1	94	19.0
Reduced variant	12.1	1	0.5	12	0.3	0	0.0
Organic variant, oxidised	12.2	2	1.0	25	0.6	0	0.0
Fine organic variant	12.21	4	2.0	30	0.7	0	0.0
Limestone tempered variant	12.4	1	0.5	21	0.5	0	0.0
Sandy oxidised ware	13	1	0.5	27	0.7	0	0.0
Fine sandy grey ware	14	22	10.9	437	10.8	100	20.2
Sandy ware	15	2	1.0	12	0.3	0	0.0
Malvernian metamorphic (H) (**MAL RE A**)	3	1	0.5	55	1.4	0	0.0
Malvernian metamorphic (W)	19	13	6.4	59	1.5	0	0.0
Worcestershire imitation BB1	149	1	0.5	7	0.2	0	0.0
Total Regional ware		*88*	*43.6*	*1740*	*43.0*	*194*	*39.3*
Southeast Dorset BB1 (**DOR BB 1**)	22	59	29.2	1033	25.6	76	15.4
Harrold Shelly ware (**HAR SH**)	23	17	8.4	232	5.7	58	11.7
Lower Nene Valley c. c. ware (**LNV CC**)	28	1	0.5	13	0.3	0	0.0
Oxfordshire red-slipped ware (**OXF RS**)	29	30	14.9	593	14.7	111	22.5
Oxford white ware (**OXF WH**)	33.1	1	0.5	191	4.7	10	2.0
Oxford parchment ware (**OXF PA**)	40	3	1.5	122	3.0	30	6.1
Total Traded ware		*111*	*55.0*	*2184*	*54.0*	*285*	*57.7*
Amphorae, Class 25 (**BATAM 2**)	42.1	1	0.5	74	1.8	0	0.0
South Gaulish Samian, Lezoux (**LEZ SA**)	43.2	2	1.0	45	1.1	15	3.0
Total Imported ware		*3*	*1.5*	*119*	*2.9*	*15*	*3.0*
TOTAL		**202**		**4043**		**494**	

TABLE 3.2 SUMMARY OF THE EVALUATION ASSEMBLAGE (WCM 101108) BY FABRIC

1993, fig. 152, 374). The plain-rimmed dishes are only broadly datable to the 2nd to 4th century, but the presence of a slip on one supported a later date (Fig. 3.6, 10). The absence of flange-rimmed bowls (*ibid*, fig. 124, type 25) may be chronologically significant. At Alcester, for example, they seem to go out of use by the end of the 4th century (Evans 1994, 146). Of particular interest was a body sherd from a jar with a repair hole (not illustrated). This sort of repair is typical of late Roman assemblages. At Wroxeter, for example, it is first noted in assemblages dated *c* AD 367–450 (Barker *et al* 1997, 240), becoming more common in contexts dated between *c* AD 500 and 660.

Severn Valley ware forms were diagnostically late types. They included three jars and a bowl (Fig. 3.6, 3–6), supporting the dating evidence from the traded wares. Two, or possibly three, bases appeared to have been deliberately rounded, perhaps for use as lids or large counters (not illustrated).

The other main local fabric was a fine, sandy grey ware (WHEAS 14), which included an everted rim jar (Fig. 3.6, 7). This fabric and another grey ware with blackened surfaces (WHEAS 149) were both similar to fabrics recovered from a late Roman site at Wyre Piddle (Laura Griffin, pers. comm.).

Catalogue (Fig. 3.6)

WHEAS 12, SVW OX 2: Severn Valley ware, plain

1 JW20.02. Hooked rim from a short-necked, Severn Valley ware, wide-mouthed jar. Dating broadly to the late 3rd to 4th century (Webster 1976, fig. 5, type C27). Diam. 250mm (12%). Well construction pit (109), context 104.

2 TC16. Splayed, plain rim from a tankard, a typical 4th-century type (*ibid*, fig. 7, E44). Diam. 170mm (8%). Ditch 106, context 105.

3 JN2.01. Small, narrow-mouthed jar with a pulley rim, a type dated by Webster to the 3rd to 4th century (*ibid*, fig. 3, A10, 11, 13; Peacock 1965–67, fig. 3, 30). Diam. 100mm (12%). Well fill context 103.

4 JN15.01. Small, short-necked jar with a near triangular rim. This type of rim occurs from the 2nd to the 4th century (Webster 1976, fig. 1, A4) but the short neck supports a later date. Similar rims can be seen on a Severn Valley ware jar and various grey ware jars from 4th-century contexts at Bays Meadow, Droitwich (Barfield 2006, fig. 100, 215, 236–38). Diam. 120mm (15%). Well fill context 103.

5 JW 7.03. Short-necked, wide-mouthed jar with a slightly squared rim. Similar to types produced at the 4th-century Hygienic Laundry site in Malvern (Peacock 1965–67, 25, fig. 3, 36–44) and found at Malvern, Newland Hopfields (Evans 2000, fig. 21, type 50; table 12). Diam. 220mm (12%). Well fill context 103.

6 BC8.04. Bowl with a grooved rim, similar to a type dated by Webster to the 3rd century (Webster 1976, fig. 9, 51, 57) but with parallels in late Roman contexts at Droitwich (Rees 1992, fig. 30, 8, 9). Diam. 140mm (20%). Well fill context 107.

WHEAS 14: Fine sandy reduced ware

7 JM7.12. Rim and base from a splayed rim jar, copying a BB1 form. The base is crudely finished, with string marks still visible. The illustrated vessel is sooted externally, and has limescale internally and below the rim externally. It is similar to a Roman vessel recovered from a post Roman context at Droitwich (Rees 1992, fig. 31, 18). Diam. 120mm (100%). Well fill contexts 103 and 107.

WHEAS 22, DOR BB 1: Southeast Dorset Black Burnished ware

8 JK7.3. Rim and joining body sherds from a BB1 jar. The form is most similar to Dorchester Greyhound Yard type 2 (Seager Smith and Davies 1993, fig. 122), rather than the characteristically late Roman, splayed rim types (*ibid*, fig. 122, type 3). However, forms similar to this are published from 4th-century or later contexts at Dorchester Greyhound Yard (*ibid*, fig. 152, 374). Decorated with diagonal pattern burnish (*ibid*, fig. 132, D9). Diam 140mm (30%). Well fill context 103.

9 JK7.3. Similar form, decorated with obtuse cross hatch pattern burnish, delineated above by a burnished groove. Diam. 170mm (37%). Well fill contexts 103 and 107.

10 DB16.01. Plain-rimmed dish. The type is only broadly datable to the 2nd to 4th century (*ibid*, type 20). However, the illustrated example appears to have been slipped, a characteristic of later examples (*ibid*). Diam. 160mm (9%). Well fill context 107.

WHEAS 23, HAR SH: Shell-tempered ware

11 JK20.01. Rim fragments, body sherds and base from a rilled jar. The rims are slightly undercut, similar to vessels published from late Roman and post Roman phases at Droitwich (Rees 1992, fig. 33, 24), and from late Roman contexts at Alcester (Lee *et al* 1994, fig. 2, R.43; fig. 6, R103, R106, R107). Diam. 150mm (58%). Well fill contexts 103 and 107.

WHEAS 29, OXF RS: Oxfordshire colour-coated ware

12 BKL1.01. Bead rim and sloping neck from a beaker. This is broadly a Young type C22 or C23, a type most common in the 4th century (Young 2000, fig. 55). Diam. 80mm (19%). Well fill context 108.

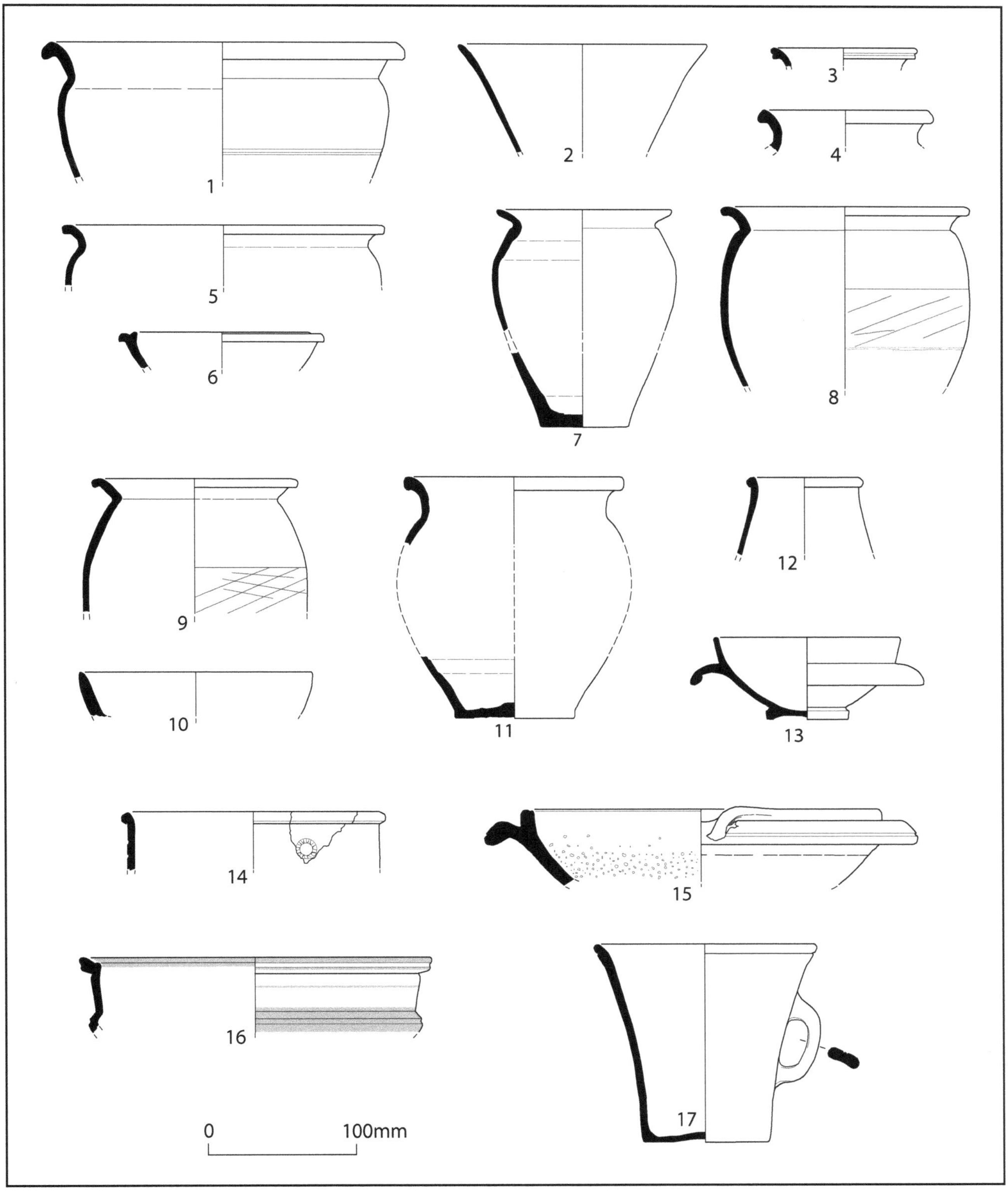

Figure 3.6 Roman pottery

13 BI16.02. Near complete flanged bowl, a type broadly dated to *c* AD 240–400+ (*ibid*, fig. 59, type C51). Diam. 130mm (84%). Well fill context 103.

14 BC1.01. Rim from a wall sided, carinated bowl with decorated with stamped rosettes; a diagnostically mid to late 4th-century type (*ibid*, 170, fig. 65, type C83/ 84). Diam. 180mm (8%). Well fill context 103.

WHEAS 33.1, OXF WH: Oxfordshire white mortarium

15 ME8. Rim and part of spout from a Young M22 mortarium, with an upstanding rim and squat flange (*ibid*, 6, fig. 23). The type first appears in the second half of the 2nd century, but from *c* AD 300 is the principal Oxfordshire mortarium produced. It is also found in late Roman contexts at Droitwich

(Rees 1992, fig. 35, 4). Diam. 300mm (10%). Well fill context 103.

WHEAS 40, OXF PA: Oxfordshire parchment ware

16 BA8.1. Wall sided, carinated bowl with red painted decoration. The form dates broadly to *c* AD 240–400+, but was increasingly popular during the 4th century (Young 2000, fig. 27, type P24). Diam. 240mm (30%). Well fill context 108.

Discussion

While the assemblage from the well is small, and must therefore be interpreted with an element of caution, it is still one of the more significant assemblages recovered in Worcester for many years. There has been coin evidence for late Roman activity elsewhere in Worcester (eg, Darlington and Evans 1992, 99) but, prior to this evaluation, only one other late Roman assemblage had been recovered (Dalwood *et al* 1997). No Worcester site to date has produced such a well stratified, sealed and tightly dated late Roman pottery assemblage. The pottery is well preserved, providing useful functional evidence. Limescale and sooting survived well, and were noted on BB1 and grey ware jars. The range of vessels, with an emphasis on cook pots, jars and colour-coated bowls, suggests a more restricted use of pottery.

The majority of the pottery comprised 'traded wares,' ie, wares produced outside the local area and traded over a wide area of the country. The proportion of traded wares is considerably higher than most other Worcester assemblages. This could partly reflect the status of the assemblage, perhaps linked with a building of some status in the vicinity. It is worth noting, however, that although Oxfordshire table wares are well represented, only the most common Oxfordshire products appear to have reached the site (Young 2000, types C51, C83/ 84, P24, M22). Evidence from the other late assemblages discussed in this volume suggests the increase in traded wares reflects a wider change in the pattern of pottery supply to Worcester during this period.

The low level of Severn Valley ware (24% by weight) is particularly significant, and is comparable with evidence from the Castle Street site (22% by weight, Dalwood *et al* 1997, appendix 4). Again, this seems to represent a general trend. Earlier assemblages are dominated by Severn Valley ware jars, and tankards (Darlington and Evans 1992, 41). The former are poorly represented in this assemblage, and the latter entirely absent. The evidence from other late sites hints at tankards having gone out of production by this period. None were included in the assemblage from the late 4th-century Severn valley ware kiln at Meole Brace, Shropshire (Evans *et al* 1999, 12), or from the contemporary Wroxeter, Tern-Severn production site (Houghton 1964), though admittedly both of these assemblages are small. Nor were any recovered from the late Roman villa site at Whitley Grange, Shropshire, not far from the Meole Brace kiln (Evans *et al* 1999). Nearby at Malvern, tankards have been found at most of the known/ suspected production sites; at the Newland Hopfield site, for example, they represented 15–17% of the vessels recovered (Evans *et al* 2000, fig. 11a and b). The late Roman, Hygienic Laundry site is a notable exception to this (Peacock 1965–67; Evans *et al* 2000, table 13). The Hygienic Laundry kiln was, in fact, producing a narrow range of jar forms similar to those recovered from The Butts. Tankards are also noted to be less frequent in the 4th-century assemblage from Bays Meadow, Droitwich (Barfield 2006, 153). This is reflected in the fact that profiles of tankards were illustrated from all Roman phases except Phase 4, and that the single body sherd illustrated from this phase was thought to be residual (*ibid*, fig. 100, 193).

A proportion of the assemblage remains preserved *in situ* in the well, with potential, perhaps, for future study, should there be further development of the site. This opportunity, should it arise, would cast more light on the character of the assemblage. A number of joins were noted across contexts; it would be interesting to know if vessels were complete when they were deposited in the well. It would also be interesting to study material from the lowest deposits, perhaps including vessels dropped in to the well when it was in use.

STONE

The most abundant building material within the excavated part of the well shaft was split hard sandstone roofing slab, from an unknown source. Few fragments had recognisable features (edging or nail holes) and only a sample of this material was retained from the estimated

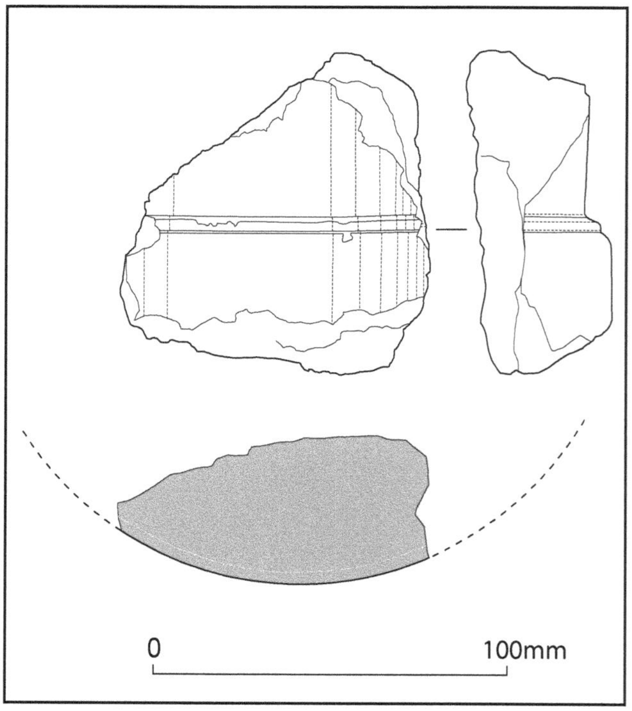

FIGURE 3.7 FRAGMENT OF OOLITIC LIMESTONE COLUMN FROM CONTEXT 108

80kg excavated. The best-preserved fragments appear to have come from hexagonal slabs with single nail holes; many of the fragments had been reddened by intense heat. A substantial number of fragments of red, brown and green local sandstone and lias were also present in the fill (samples only retained). Occasional fragments of oolitic white limestone were also present, and this was possibly sourced from the Cotswolds. Only one intact worked stone block was recovered, and this was a curved sandstone ashlar derived from the upper part of the well structure. A small fragment of oolitic Cotswold limestone column, approximately 190mm in diameter, was recovered and this showed clear signs of scorching, apparently whilst still *in situ* as only the face was affected (Fig. 3.7). This fragment is noteworthy as the first excavated evidence of a 'classical' Roman structure from Worcester.

TILE AND BRICK

The recovered tile, although a small sample, appeared to represent the full range of Roman ceramic building materials commonly encountered in Worcester, including *tegulae*, *imbrex*, (both roof tile forms) *lydion* (long flat bricks used primarily for walling), *pilae* and box flue tiles (both indicative of under-floor heating systems). All the tile material was fragmented, but cross-joining sherds representing substantial portions of individual tiles were present. One single box tile fragment represented approximately half a tile. Another box tile fragment was tapered in two directions towards the combed face – this tile from context 102 appears to have been a hollow voussoir from a dome or vault. Some of the flat tile was also scored and had mortar adhering, suggesting that it had also been used to create wall flues; joining fragments of the same scored flat tile were present in contexts 103 and 108. The tiles were of varying fabric and form (Fig. 3.8), suggesting that they represent material from several phases of construction, or at least varying suppliers. Of particular interest was the presence of two tile fragments, both apparently reused, of different forms and slightly different fabrics both bearing the same hobnailed sandal or boot imprint (Fig. 3.9). These tiles clearly came from the same workshop producing both *tegulae* and pierced flat tile, and it is a remarkable coincidence that both should have been disposed of in the same feature.

The nature of the tile assemblage was such that it appears to have been material rejected as too fragmented for further re-use, rather than demolition material dumped *en masse*. This is important, as it suggests that there were still masonry buildings being constructed locally at the end of the 4th century. Previous re-use of the materials was evidenced by *tegula* fragments that had been trimmed of their flanges and had mortar stains across the breaks. As the assemblage was comparatively small, no truly statistically valid interpretation of the proportional presence of the various tiles can be deduced, but it would appear that the material present represents a cross sample of the demolition debris from a whole masonry building rather than simply re-roofing or alterations to walls or floors. *Tegulae* and *lydion* were present in equal proportions by weight (both 30% in context 108) followed by box tile (19.08%) and *pilae* (12.93%); *imbrex* appeared to be under represented, but this may reflect re-use of *tegulae* as walling material. This represents only a very crude comparison as the various tile forms vary significantly in weight; a single *lydion* might weigh twice as much as a *tegula* and three times as much as a box tile.

TESSERAE

The number of *tesserae* (almost 200 recovered from such a small excavated sample) was remarkable as they are not frequently found in the city. Almost all of the *tesserae* appeared to be light grey-brown lias, a significant proportion of which had been subjected to heat and had fissured as a result. The remaining *tesserae* were of cut tile and possibly derive from a border. The stone *tesserae* were crudely cut and appear to have been part of a tessellated pavement or floor rather than a mosaic. Very few retained any trace of mortar bedding. The *tesserae* were distributed fairly evenly throughout fill context 103 and had not been dumped as a single deposit, perhaps scraped up inadvertently with other demolition material and rubbish which had accumulated on a decaying tessellated surface. Only a small proportion of the *tesserae* came from the deepest excavated context (107).

PLASTER AND MORTAR

There was approximately 4.5kg of wall plaster recovered from the fills of the well, representing the largest assemblage of painted plaster from a single deposit yet discovered in the city. Much of the plaster was heavily fragmented and weathered, but sufficient traces of paint survived to indicate that the decorative scheme was predominantly deep red and white, enlivened by panels edged in yellow, small areas of pink and rare black stripes. The plaster was predominantly mortar backed and most appeared to have come from smooth masonry walls, though some timber impressions were also noted. The mortar was of a consistent light yellowish-brown sandy mix and included moderately frequent grits. The fragments of wall plaster recovered appear to have been heavily weathered prior to deposition; very few fragments retained paint across their whole surfaces, and many were pitted and scored. Breaks in the mortar backing, by contrast, appeared to be fresh at the time of deposition; this suggests that the plaster was stripped from walls that had been left exposed to the elements or to wear and tear for a considerable period prior to demolition. None of the plaster showed signs of burning (in contrast with the roofing slabs and stone column fragment which had been subjected to scorching). Small occasional fragments of pinkish mortar containing tile and lime flecks (*opus signinum*) were also noted; one tile fragment also retained traces of *opus signinum*.

COAL

The presence of several fragments of coal in the well is of considerable significance as it indicates a hitherto rarely

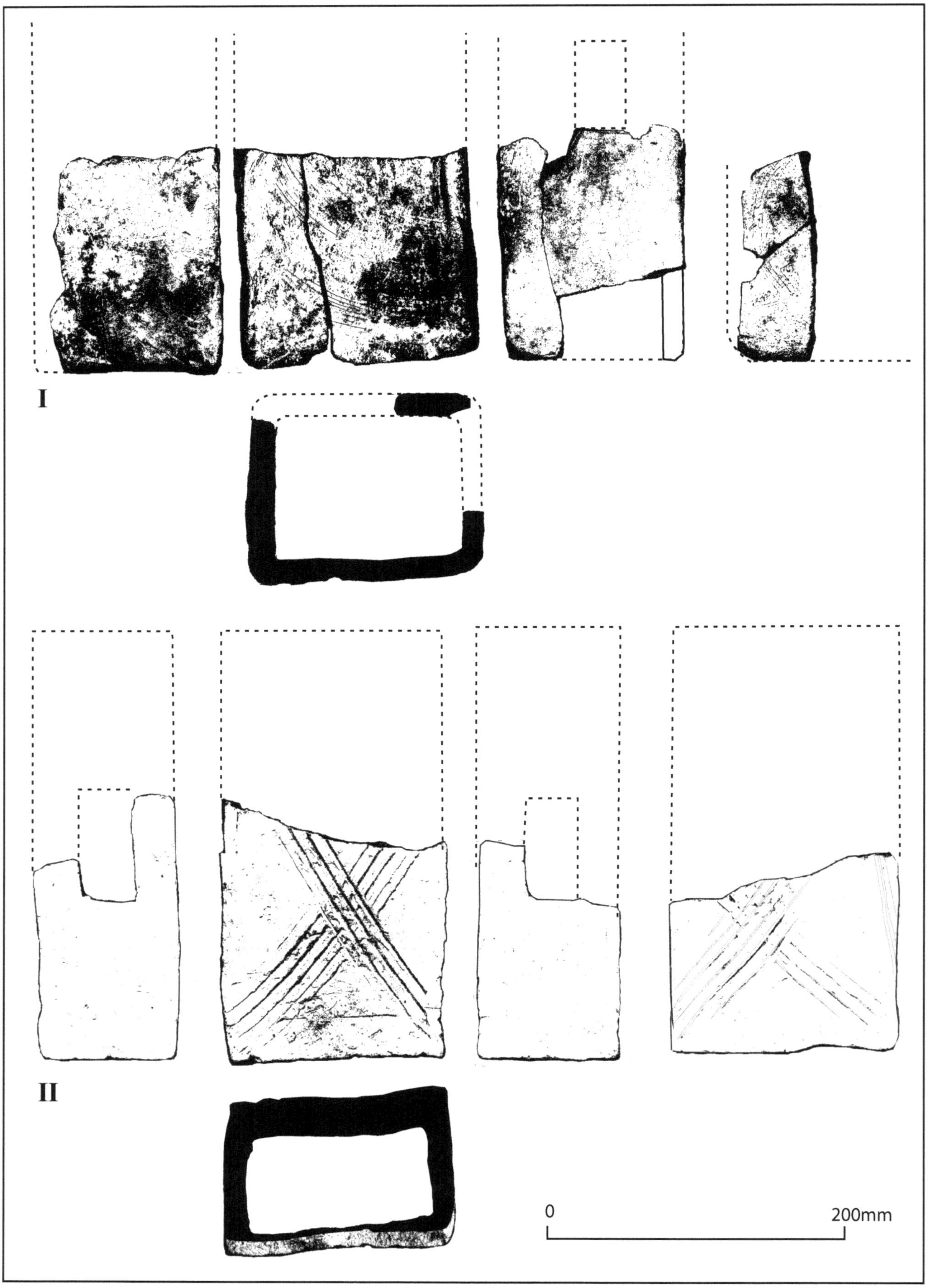

Figure 3.8 Box tile from context 108; I – five adjoining fragments in (overfired?) sandy brown fabric; II – fragment in hard orangey red fabric

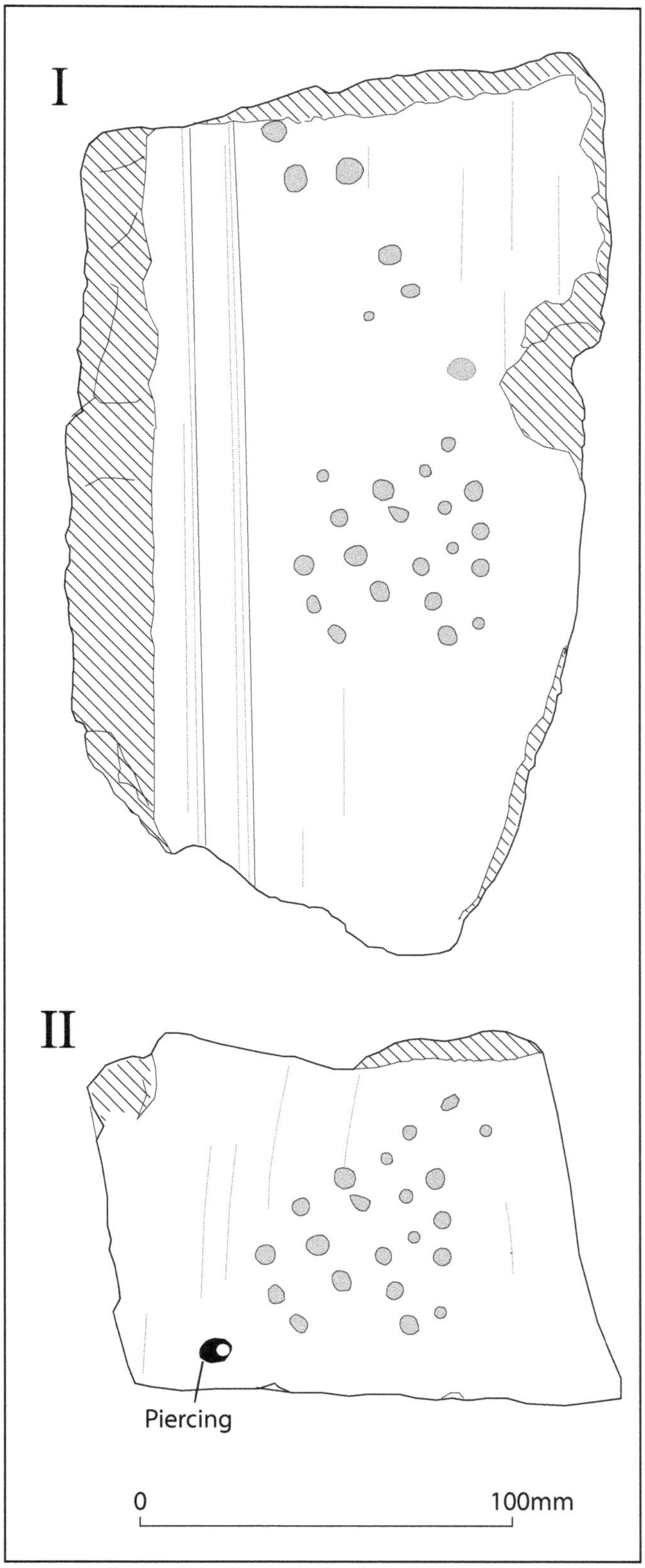

FIGURE 3.9 HOBNAIL IMPRESSIONS ON ROMAN TILES; I – IMPRESSION ON *TEGULA*; II – IMPRESSION ON PIERCED FLAT TILE. BOTH IMPRESSIONS ARE FROM THE SAME BOOT OR SANDAL.

documented source of fuel for Worcester in the Roman period. Whilst coal was not useable as a smelting fuel for ironworking prior to the introduction of the blast furnace, it would have served as a general heating fuel and may have been sourced and shipped from shallow mines in the Forest of Dean together with iron ore. Shallow coal was also available more locally in the Pensax area, but it is not known if these resources were utilised in Roman times. Small fragments of charcoal were also present in the fills of the well, identification of which indicated that they included the usual range of native fuel woods, and may have derived either from domestic or industrial processes.

OTHER EVALUATION AND WATCHING BRIEF CONTEXTS

GLASS

The glass from context 202 consisted of three small joining sherds (originally a single sherd, but crushed *in situ*) from a blue and yellow 'snake-thread' decorated, very fine colourless vessel – these are generally attributed to the late 2nd and 3rd centuries, and a source in the Cologne area has been postulated (Allen 1998). The rejoined fragment illustrated in Figure 3.10 appears to be a stylised leaf design.

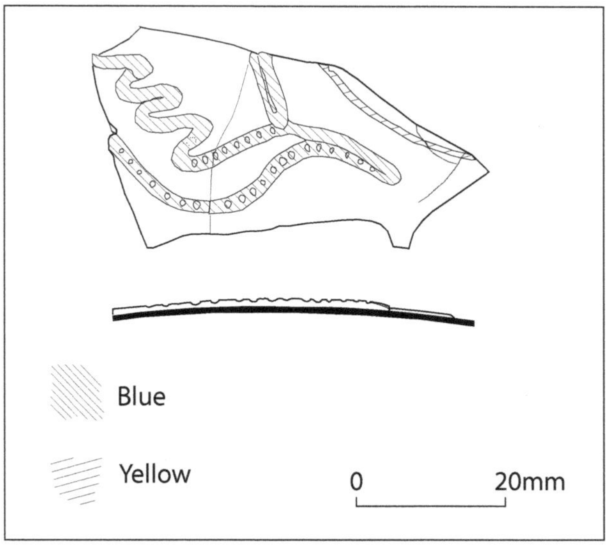

FIGURE 3.10 SNAKE THREAD GLASS FROM CONTEXT 202

ROMAN POTTERY *C Jane Evans*

Introduction

Ditch 106, recorded during the evaluation, produced nine sherds of Roman pottery and a further ten were unstratified (Table 3.1). The pottery from the ditch included a typically 4th-century Severn Valley ware tankard (Fig. 3.6, 2), and characteristically 4th-century fabrics such as Oxfordshire colour-coated ware and shell tempered ware. The watching brief produced 75 sherds of pottery (Table 3.1), comprising a near complete Severn Valley tankard (Fig. 3.6, 17 and Table 3.3), the complete base of a Severn Valley storage jar, and various other more fragmentary sherds. The

Common Name (National Fabric Reference code)	WHEAS Fabric Code	Qty	% Qty	Wt (g)	% Wt	Rim %	% Rim EVE
Severn Valley ware, oxidised (**SVWOX 2**)	12	33	44.0	2532	77.2	98	83.8
Reduced variant	12.1	2	2.7	11	0.3	0	0.0
Fine organic variant	12.21	1	1.3	18	0.5	0	0.0
Reduced organic variant	12.3	4	5.3	22	0.7	0	0.0
Malvernian metamorphic (H) (**MAL RE A**)	3	6	8.0	126	3.8	0	0.0
Total Regional ware		*46*	*61.3*	*2709*	*82.6*	*98*	*83.8*
Southeast Dorset BB1 (**DOR BB 1**)	22	22	29.3	126	3.8	14	12.0
Oxford white ware (**OXF WH**)	38	1	1.3	1	0.0	0	0.0
Total Traded ware	*60*	*23*	*30.7*	*127*	*3.9*	*14*	*12.0*
Amphorae, Class 25 (**BATAM 2**)	42.1	4	5.3	247	7.5	0	0.0
Samian	43	2	2.7	18	0.5	5	4.3
Total Imported ware		*6*	*8.0*	*265*	*8.1*	*5*	*4.3*
TOTAL POTTERY		**75**		**3278**		**117**	

TABLE 3.3 SUMMARY OF THE WATCHING BRIEF ASSEMBLAGE (WCM 101109) BY FABRIC

Height (mms)	Rim Diameter (mms)	Rim Radius (mms)	Base Radius (mms)
120	160	80	42.5

TABLE 3.4 MEASUREMENTS OF THE NEAR COMPLETE TANKARD

presence of complete/ substantial fragments is reflected in the average sherd weight for the assemblage (44g).

The tankard (Fig. 3.6, 17) was a typically 4th-century type. Where complete tankards have been recovered elsewhere, there seems to be a proportional relationship between rim diameter and vessel height (Lee *et al* 1994, 10). This was illustrated graphically for the vessels from the Newland Hopfields kiln in Malvern (Evans *et al* 2000, 22–24, fig. 16). The proportions of the tankard from The Butts (Table 3.4) have a similar ratio. The only other Severn Valley ware form was a fragmentary hook-rimmed jar (Webster 1976, fig. 5, 25–28), dating broadly to 2nd to 4th century. BB1 forms included a plain-rimmed dish, not closely dated (Seager Smith and Davies 1993, type 20), and a fragmentary jar rim.

Further evidence for late activity came from sherds of Black Burnished ware (BB1) with obtuse cross hatch burnish, characteristic of *c* AD 280–370. There were also sherds of coarse Malvernian ware, similar to the 'oven/ beehive' fragments noted at 8–12 The Butts (see Evans, Chapter 4).

Some residual fabrics were evident: organic Severn Valley ware (WHEAS 12.21, 12.3), Dressel 20 amphora (WHEAS 42.1) and samian ware (WHEAS 43), the latter possibly East Gaulish samian from Rheinzabern.

Catalogue

WHEAS 12, SVW OX 2: Severn Valley ware, plain (Fig. 3.6, 17)

17 TC16. Near complete tankard with splayed walls and a grooved bead rim. The general type is dated to the 4th century by Webster (1976, fig. 7, E44), though similar examples are illustrated from Phase 3c at Droitwich, Bays Meadow (Barfield 2006, fig. 98, 138, 140). The end of this phase is dated on coin evidence to the very late 3rd century, *c* AD 293–96 (*ibid*, 125). The tankard from the watching brief is slightly distorted; the rim is not quite round and the height varied around the rim. Diam. 160?mm (90%). Context 202.

STONE

Occasional Cotswold stone roofing slab fragments were noted on the surface in the area of the well, but none were securely stratified. Other small fragments of sandstone were noted during the groundworks, but these may all be attributed to the construction and rebuilding of the adjacent city wall.

ENVIRONMENTAL EVIDENCE

SOIL SAMPLE FROM WELL FILL CONTEXT 103 *Elizabeth Pearson*

An environmental sample of approximately 40 l of soil from context 103 was processed by flotation followed by wet sieving using a Siraf tank. The flot was collected on a 300 micron sieve and the residue retained on a 1mm mesh, allowing for the recovery of items such as small animal bones, molluscs and seeds. The residue was fully sorted by eye and the abundance of each category of environmental remains estimated. The flot was fully sorted using a low power EMT stereo light microscope and plant remains identified using modern reference collections maintained by the Service, and a seed identification manual (Beijerinck 1947). Nomenclature for the plant remains follows the Flora of the British Isles, 3rd edition (Clapham *et al* 1989).

This deposit was rich in finely fragmented charcoal, large mammal bone fragments (some burnt) and one fragment of wild/ dwarf cherry stone (*Prunus avium/ cerasus*). The latter is likely to be the remains of food waste disposed of in the well, although it is possible that wild or cultivated trees were growing within the vicinity, perhaps in an orchard or garden. The only cereal crop debris recorded was one grain of charred emmer or spelt wheat (*Triticum dicoccum/ spelta*). Mortar fragments were also moderately abundant and one fragment of sandstone was recovered from the sample. The assemblage is likely to represent waste from a hearth and the demolition of a building.

Preservation of environmental remains was poor, with only fragmented evidence of waste disposal on the site during the late Roman period. However, there is potential for samples at greater depth in the well to be waterlogged and hence in these conditions, survival of organic remains may be good.

DISCUSSION

The presence of a stone-built well of Roman date is of some interest, as few have been previously identified in the city. This well is also of particular interest in that its structure included at least one reused stone moulding from an earlier building (and may include considerably more as the structure was not examined in any detail nor dismantled). Identifiable fragments of architectural stone mouldings of Roman date are extremely rare finds in the city. Added to this important evidence for a stone building pre-dating the well itself, was the large volume of building material recovered from the infill of the well. This latter material presumably represents the demolition of a second masonry building of some pretensions in the late 4th century, quite probably between AD 370 and 390.

The discovery of evidence for a substantial and apparently public or high status Roman building in this area was unexpected, as it contrasts strongly with the previous observations of low-level Roman industrial activity in the areas to the south at Blackfriars and to the northeast at Farrier Street. Extensive observations at Blackfriars (Barker 1968–69 and Mundy 1986) produced only very small quantities of Roman building materials, and a similar decline in density of excavated building materials such as tile and wall plaster has been noted on the sites to the north (8–12 The Butts, see Chapter 4) and at Farrier Street (Dalwood *et al* 1994). The very small-scale observation within Tramps nightclub (Napthan 2002), immediately to the southeast of the present site, did, however, produce a comparatively high proportion of tile and ceramics, and it is probable that the building or buildings from which the debris in the well derived was/ were located in the area immediately adjacent to the well, possibly in the area of the Angel Centre and the former Congregational Church, now Tramps nightclub. Parts of the Roman building or complex may have been removed by digging of the medieval city ditch, but elsewhere structural remains may survive. An observation of sandstone rubble in Angel Place during drainage works in 1991 (Worcester City HER files)[1] might perhaps also relate to the same building.

The precise function of the building remains uncertain, but the presence of large quantities of painted wall plaster and box tile together with hypocaust stack bricks, *tesserae* and a column fragment strongly suggest a public or high status building. The latter is perhaps unlikely given the proximity of the contemporary industrial activity, and a public function is therefore the most probable. Springs were utilised in the immediate area during the medieval and post medieval period (the 'Holy Well' was immediately adjacent, and Prides' Cold Baths close by), and it is not unlikely that a Roman bath house should have stood in this area. Little firm evidence was recovered, though one tile fragment (context 108) appears to have been from a hollow voussoir tapered to fit in a vault or dome.

The fabric of the well also included stone that had clearly been reused from an earlier high status building. Given that the stonework was apparently reused in the 3rd century, it might well derive from a building of the 2nd century or earlier. Naturally, there is a strong possibility that the stone was brought to the present site specifically for the purposes of constructing the well, and the earlier building may have lain elsewhere. Knowledge of the masonry buildings of Roman Worcester is extremely rudimentary, and the only high-status building presently confirmed is that in the Britannia Square/ Moor Street area, known almost entirely from observation of residual building materials (Napthan 2002) and antiquarian reports, rather than direct observation. Only further excavation may clarify the precise nature and location of these buildings, but the present project, together with recent work elsewhere in the Butts, does tend to confirm that the Roman settlement lay largely to the north of the medieval city and had several informal foci rather than a planned urban centre.

The excavated material within the well consisted of a comparatively dense concentration of artefacts (estimated at 3–5% of the well backfill by weight), but has no characteristics that might suggest that the selection of materials was in any way deliberate. The range of material, level of fragmentation, the presence of cross-joining sherds and the incompleteness of the discarded objects were all typical of an urban refuse deposit. Very little of the material seems to have been residual. The oldest objects represented were perhaps the fragments of glass vessels, but these were no doubt valued items and,

[1] Site notes made by Charles Mundy, former City Archaeologist, during the Black Friars Project

barring accidents, were retained across generations. The presence of a significant number of nails throws some light on the quantity of organic materials (wood, leather, etc) originally discarded in the well but now almost invisible in the archaeological record. It is entirely possible that such organic materials do survive at the base of the shaft, where anaerobic conditions are likely to exist below the water table.

The presence of the well within the postulated Roman road line across this area adds to the growing body of evidence that the road line terminated in the Blackfriars area. No sign of the road was seen in the apparently Roman cultivated soils observed in section beneath the medieval city wall line and, as reported elsewhere in this volume, the road was not seen on either 8–12 The Butts (see Chapter 4) or 14–24 The Butts (see Chapter 5). The metalling at Farrier Street lay on the approximate line of the postulated road, but was without any conclusive features, and might also be interpreted as a yard surface.

The remaining unexcavated fills are likely to contain a considerable body of evidence relating to the demolished buildings. A broadly similar stone-lined well, subsequently discovered at 14–24 The Butts, was examined by augering and found to be 6.5m deep (see Chapter 5) – this would suggest that at least 5m of unexcavated deposits may remain at 1 The Butts.

CONCLUSION

Preservation of Roman deposits was restricted almost entirely to cut features within the site boundaries, though Roman cultivated horizons appear to survive to a height of approximately 20.45m AOD beneath the city wall. The medieval city ditch had removed all earlier deposits across the majority of the site, leaving a small area of berm between the ditch and city wall in which Roman deposits appear to have survived to within about 0.5m of Roman ground levels (as projected from neighbouring sites, eg, 20.29m AOD at 8–12 The Butts and 20.75m AOD within the Tramps building). Whilst truncated, the Roman deposits proved to be artefactually rich and included a regionally significant assemblage of late Roman pottery and building materials from the well.

Little was seen of further Roman features, except pit 201, which also appears to have been exceptionally artefactually rich if the small sample excavated was representative. Much of the Roman stratigraphy was truncated by the creation of the medieval defences. These were represented by the fragmentary and largely re-built or refaced section of city wall, together with a wide area of deep post-medieval deposits filling the line of the medieval ditch. The fills of the ditch were not assessed in detail during the current project, but elsewhere (including the adjacent site at 3–5 The Butts) have proven to include archaeologically significant deposits of medieval and post-medieval date.

ACKNOWLEDGEMENTS

This project was carried out between May 2003 and January 2004 by Mike Napthan Archaeology on behalf of the site owners (Tramps nightclub). The assistance and cooperation of the following was much appreciated: the architects Martin and Nicolette Baines, the client Mr Derek Holder and his building contractors Wilden Construction. Kerry Whitehouse assisted with the fieldwork and post-excavation processing. Specialist reports were contributed by C Jane Evans and Liz Pearson with coin identifications by Angie Bolton. The project was monitored on behalf of Worcester City Council by James Dinn (Archaeology Officer). Copy editing was carried out by Malcolm Hislop.

CHAPTER 4: 8–12 THE BUTTS

Mike Napthan

with C Jane Evans and Elizabeth Pearson

SUMMARY

The site at 8–12 The Butts had recently been cleared of a number of late 18th- to early 19th-century workshop buildings and dwellings in advance of residential redevelopment. Archaeological evaluation in 2003 involved the excavation of two trenches, and the removal of a section of cellar wall, to reveal the overall depth of deposits. The northernmost trench exposed, at its eastern end, a complex and apparently undisturbed sequence of Roman and immediately post-Roman deposits. Features included gravelled surfaces, pits and a stake-built structure. The earliest occupation appeared to date from the late 1st to early 2nd century, and consisted of a large pit sealed by gravel surfaces of similar date. Subsequent surfaces appeared to be of 2nd- to 3rd-century date, but continued occupation of the area into the 4th century was attested by material residual in post-medieval contexts.

Later Roman or post-Roman 'dark earth' deposits survived to within a few inches of the floor levels of the recently demolished building. The central area of the northern trench contained a very large, and apparently linear, post-medieval feature, probably a Civil War defensive ditch, although post-medieval artefacts were very sparse within the fills. The alignment of this feature was unclear, but probably ran southwest–northeast. To the west of this feature a small area of gravelled surface of Roman date survived, but had been cut away to the west by a complex of deeply cut post-medieval features, possibly sandpits, which were not investigated in detail.

The Roman and immediately post-Roman deposits appeared to be well preserved in the northeastern half of the site, albeit locally truncated by post-medieval features. In contrast, deposits in the trench excavated on the southwestern side of the property had been affected by extensive pit digging and the possible terracing down of The Butts frontage to below the top of natural deposits. The cellared areas on The Butts frontage appeared to retain no significant deposits.

INTRODUCTION

This chapter represents a summary of the findings of an archaeological evaluation undertaken at 8–12 The Butts prior to redevelopment of the site. The site (NGR SO 8478 5512) lies in the historic core of Worcester City (Archaeologically Sensitive Area no. 18) and the Historic City Conservation Area (Figs 1.2, 1.3 and 4.1). The project was designed to meet a brief prepared by Worcester City Council's Archaeology Officer, and was undertaken in accordance with the *Standard and Guidance for Archaeological Evaluation* issued by the Institute of Field Archaeologists (1994) and *Archaeological Guidance Paper 4: Archaeological Watching Briefs: (guidelines)* issued by English Heritage. The evaluation of the site was registered on the Worcester City Sites and Monuments record as WCM101173.

The road line of the Lower Butts appears to have had post-medieval origins, and the area to the north remained largely as 'tenter' grounds on which cloth was dried after dyeing in the 17th and 18th centuries. The present property fell within the Netherton Estate, centred on a large post-medieval house to the west. Post-medieval activity on the site included stables and a shoeing smithy established by the early 19th century, at which time much of the Netherton Estate, consisting of small yards surrounded by workshops and some living accommodation, was leased out for similar equestrian-related functions. More recent activity on the site has included a mid 20th-century upholsterer's workshop, and an antique dealer's workshop. There was also some evidence of motor-trade activity on the site.

METHODOLOGY

As shown in Fig. 4.2, two evaluation trenches were excavated, using a mechanical excavator, in locations designed to expose the line of the postulated Roman road from Blackfriars. A section of wall and an area of floor were removed from the former cellar adjoining the Paul Pry Public House to determine the likely vertical extent of deposits. All subsequent excavation and cleaning was undertaken by hand.

RESULTS

The natural subsoil consisted of firm yellow and reddish brown sand, with clear gravel lenses following the bedding planes, observed in the section behind the cellar wall. The natural deposits were within 0.3m of current ground levels at their highest point (20.20m AOD) which was in the area immediately adjacent to the eastern cellar. This compares fairly closely to the highest observed natural at 1 The Butts (20.49m AOD, see Chapter 3) and suggests strongly

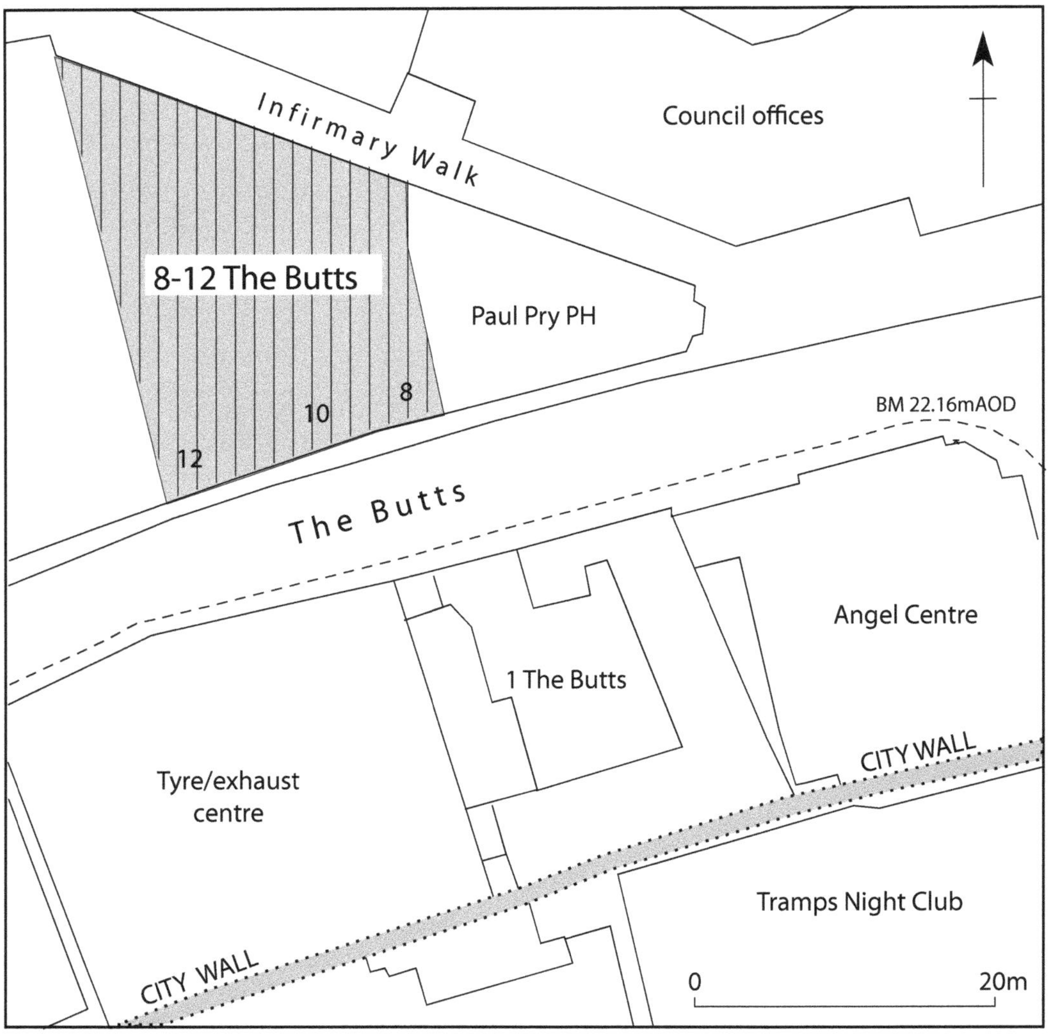

FIGURE 4.1 LOCATION PLAN

that previous reports of a natural valley in this area were either mistaken, or that the valley head terminated below this point. The natural sand deposits exposed in Trench 2 were very compact and resembled very gritty and friable red decayed sandstone. In Trench 1 the natural deposits consisted of light brown sand with fine gravel lenses.

1st- to 2nd-century Roman deposits

In Trench 1 a large pit (122) was partially exposed in plan (Fig. 4.3) and section (Fig. 4.4). It contained contexts 121, 123 and 125. No original edges of the pit were observed as it had been partly truncated by later features, but the tip lines within the feature suggested a circular or oval feature approximately 3m in diameter and over 1.5m deep from Roman ground level. The earliest excavated deposit was, the charcoal-rich grey sandy fill (123) of pit 122, with abundant animal bone, frequent Roman pottery of mid 1st- to early 2nd-century date, moderate burnt clay and occasional burnt cobbles. A small quantity of ceramic building material was recovered from pit 122 including two fragments of *pilae* or *lydion* from context 123. This fill was subjected to environmental sampling (see below). Overlying 123 was a tip of compact brown sand (125) with very few inclusions. An almost complete small-horned cattle skull was recovered from 121, but this context otherwise contained only sparse artefacts. The pottery dating for this pit, together with a probable Claudian coin, indicates a late 1st- to early 2nd-century date for the fills.

Pit 122 was sealed by at least two distinct layers of gravelling; layer 117 of 1st- to 2nd-century date and layer 116, which was undated. Both were overlain by 114, which consisted of a concreted mass of iron slag and greenish sand containing occasional mid 2nd-century pottery. This context was not exposed in plan but did not appear to be a finished or level surface. To the west, a further area of fine gravelling (118) was of similar date to 117, and both were at a similar level, suggesting that they formed parts of a larger gravelled area. The contrasting density and size of metalling may represent differential use.

Later Roman/ post-Roman deposits

Layer 112 consisted of a dark grey sandy loam, perhaps best described as 'dark earth', which was clearly cut by stakeholes and had lenses of compacted brown sand on its surface (Figs 4.3, and 4.4). The dating evidence for

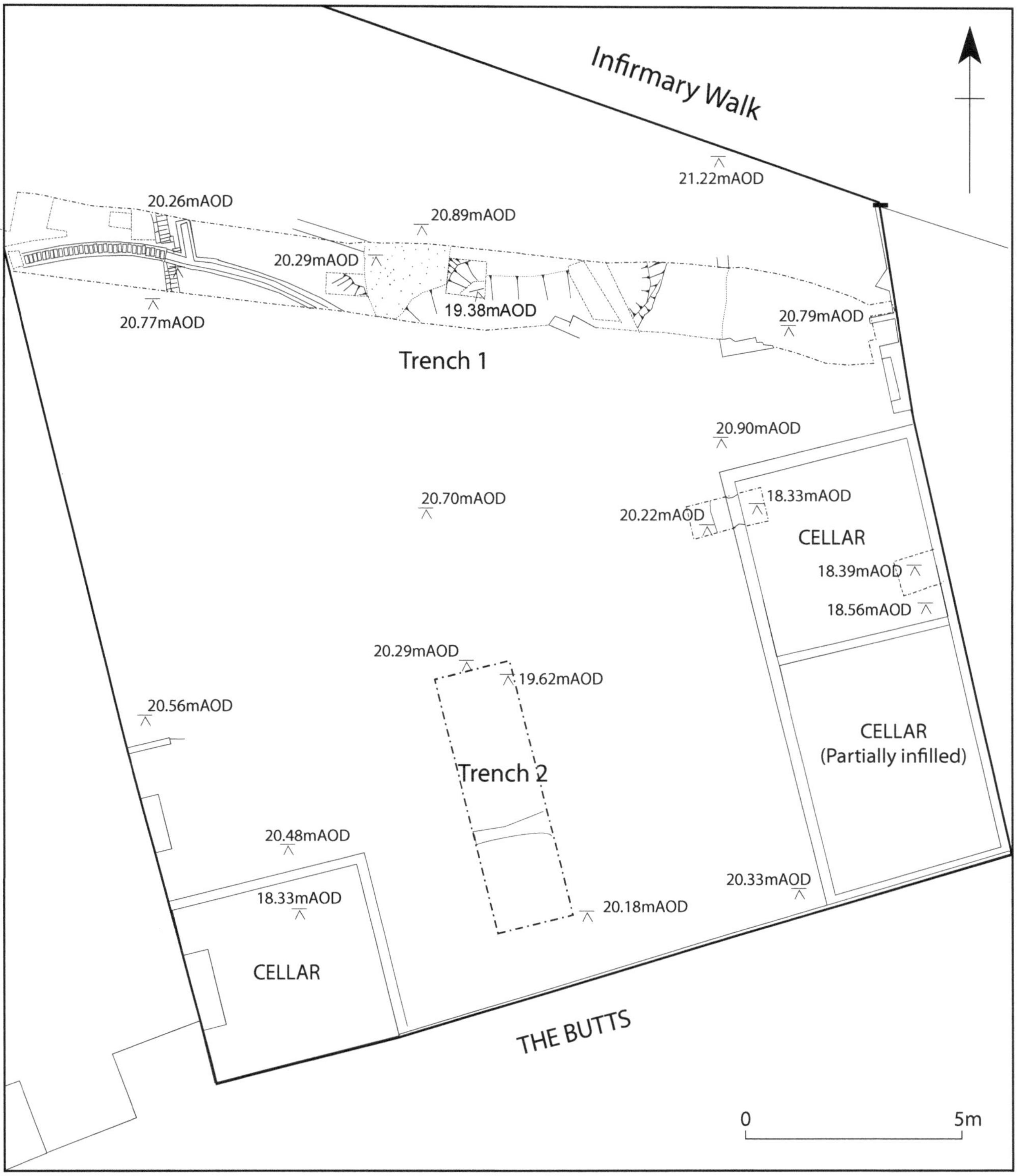

Figure 4.2 General site plan

this material was all 3rd century, but there was also a high level of residual earlier material and it is possible that all the Roman material was residual in a post-Roman soil accumulation. An insufficient area was excavated to determine any patterning in the stakeholes apparently cut into 112 – they might represent any form of stake built structure. Layers 110 and 111 were similar 'dark earth' layers and represent arbitrary spits of a single deposit of apparently cultivated soils. Small quantities of post-medieval material were recovered, but these appeared to be intrusive from adjacent cut features. The compact nature of these soils was evidence of trampling, and the presence of small fragments of slag might represent residual surface metalling.

Undated (possible Roman) deposits

A layer of sandy brown soil containing occasional charcoal and burnt clay flecks was observed in section and plan in a small hand dug test slot cut through the cellar wall

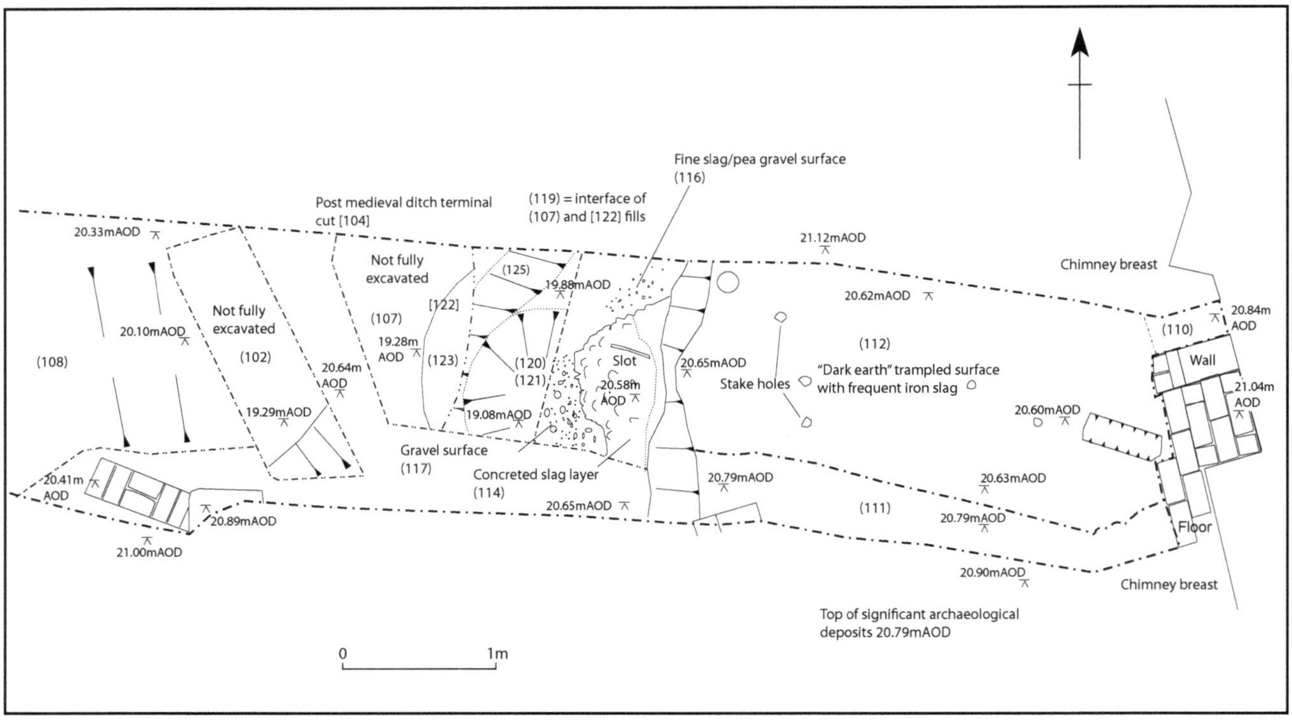

FIGURE 4.3 DETAIL PLAN OF THE EASTERN END OF TRENCH 1

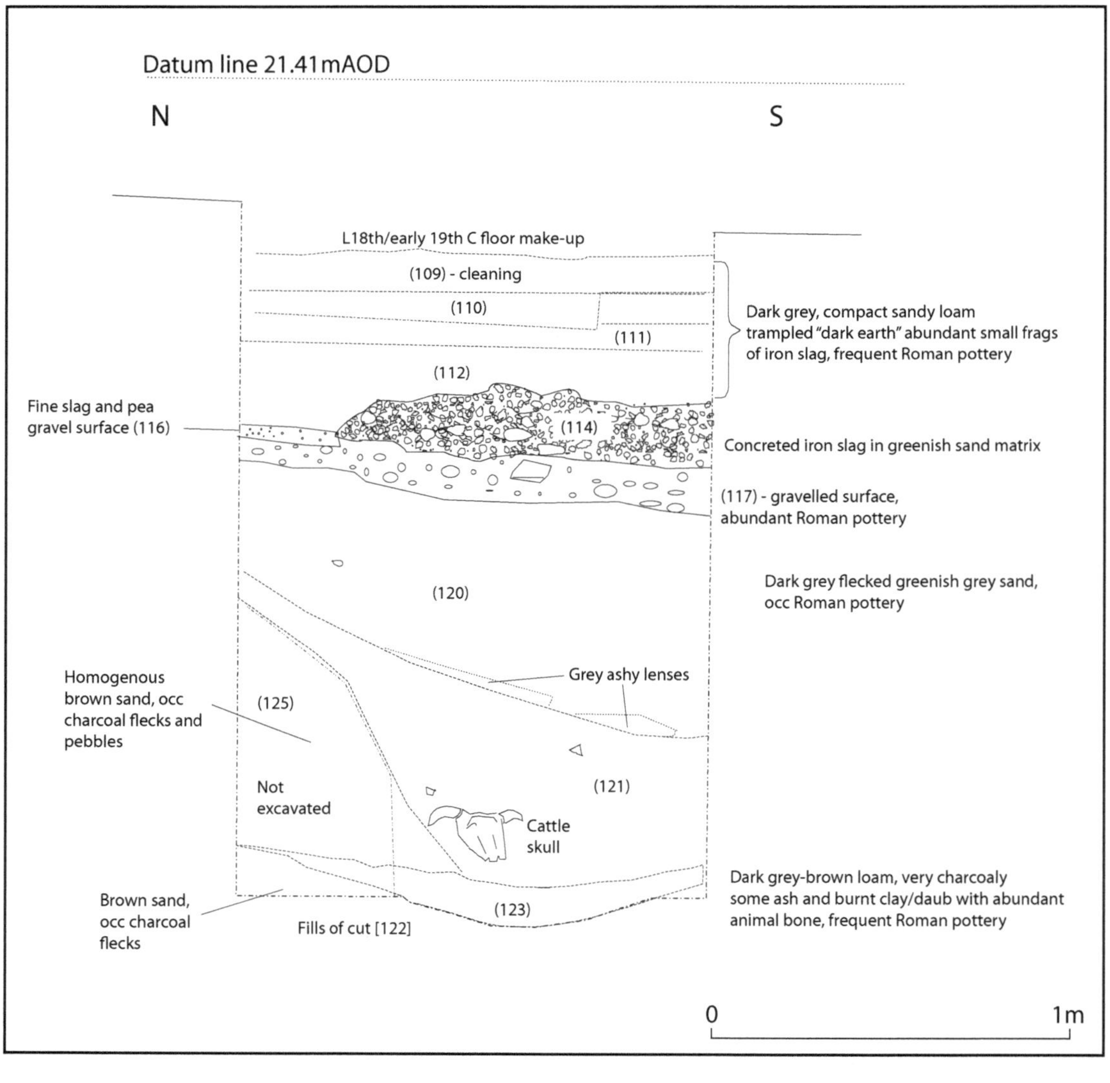

FIGURE 4.4 WEST FACING SECTION ACROSS PIT 122

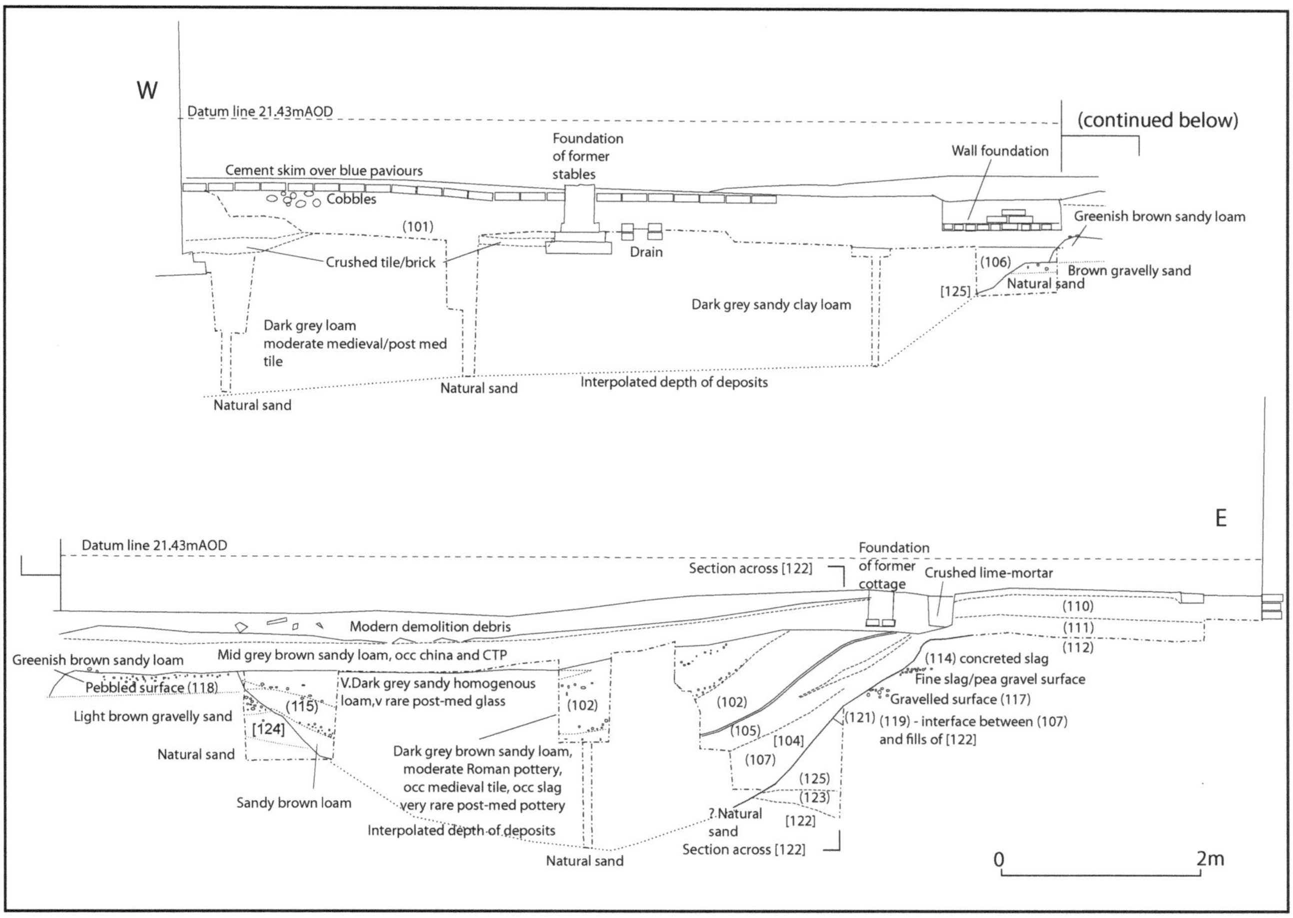

FIGURE 4.5 LONG SECTION OF TRENCH 1, SOUTH FACING

southeast of Trench 1. The nature of the material and level corresponded with Roman deposits to the north, and it may be reasonably assumed that this layer was also Roman. Significant quantities of residual Roman pottery were present on the site, and this covered a date range including the 4th century. There was however a noticeable paucity of Roman building material, no diagnostic material being recovered from later contexts.

MEDIEVAL DEPOSITS

The nature of medieval activity in the area was unclear, and no certainly medieval deposits were identified. There was, however, a small but significant quantity of residual medieval material, including a number of glazed and unglazed roof tile wasters and fragments of glazed roof and decorated floor tile. The floor tile was clearly part of a multi-tile pattern, but was too damaged and worn to be precisely identified. Given the proximity of Blackfriars, it is probable that the medieval material largely derived from activity associated with the monastery. The range of material covered the 13th to 16th centuries, and almost all were small and abraded sherds. It is perhaps noteworthy that medieval material was more abundant on this site than reported from nearby observations to the north and west.

POST-MEDIEVAL DEPOSITS

The principal post-medieval feature excavated was cut 104 in Trench 1 which appeared to be a large ditch aligned approximately southwest–northeast, possibly curving to the north beyond the trench (Fig. 4.5). Due to the size and position of this feature, it was not possible to determine its extent or alignment with any precision; however it appeared to be approximately 4m across and 2–2.5m deep. It would seem probable that the feature originated as a Civil War defensive earthwork, and its upcast bank doubtless helped preserve the Roman deposits immediately to the east.

The fills of this ditch are shown in section in Fig. 4.5. The earliest excavated fill (107) contained rare fragments of medieval tile and very rare post-medieval pottery (possible 17th century). The subsequent fills were largely devoid of artefacts and there was some evidence of a turf line at the eastern end of the feature, suggesting that 104 was partially filled then abandoned. Subsequent fills appear to have been primarily topsoil, perhaps derived from levelled earthworks.

At the western end of Trench 1 there were a number of large intercutting pits that were not examined in detail.

Those which were sampled and augered proved to contain homogenous grey/ dark grey silty loams containing frequent medieval roof tile fragments and occasional post-medieval pottery. Pit 124 additionally contained abundant small fragments of iron slag. Augering of the inter-cutting pits in this area indicated that they were up to 2.3m deep. Trench 2 contained similar inter-cutting post-medieval features, but these appeared to be shallow and of later date, relating to 19th-century construction and 20th-century demolition of buildings on The Butts street frontage. There were several indications that ground levels had been reduced along the street frontage and this may account for the apparent shallowness of deposits on the southern part of the site.

ROMAN POTTERY *C Jane Evans*

SUMMARY

The excavation produced 453 sherds of pottery, including a good early Roman group from a sealed pit, and a later Roman group associated with the dark earth layer.

METHODOLOGY

All hand retrieved pottery was examined. The comparatively small assemblage of medieval and post-medieval pottery was spot-dated but not studied in detail. Roman pottery was identified, quantified and dated to

Description	Qty	% Qty	Wt (g)	% Wt	Av Wt (g)	% Rim	% Rim EVE
Ditch 104	92	20.3	1134	19.9	12	56	11.4
Pit 122	165	36.4	2171	38.0	13	287	58.6
Pit? 124	3	0.7	32	0.6	11	12	2.4
Dark Earth	141	31.1	1630	28.5	12	107	21.8
Layers and surfaces	42	9.3	639	11.2	15	17	3.5
TR1/ TR2	10	2.2	105	1.8	11	11	2.2
TOTAL	**453**		**5711**		**13**	**490**	

TABLE 4.1 SUMMARY OF ROMAN POTTERY BY FEATURE

Common Name (National Fabric Reference Code)	WHEAS Fabric Code	Qty	% Qty	Wt (g)	% Wt	% Rim	% Rim EVE
Severn Valley ware, oxidised (**SVWOX 2**)	12	173	38.2	1645	28.8	85	17.3
Reduced variant	12.1	13	2.9	270	4.7	4	0.8
Organic variant, oxidised	12.2	62	13.7	957	16.8	66	13.5
Fine organic variant	12.21	14	3.1	420	7.4	17	3.5
Reduced organic variant	12.3	58	12.8	587	10.3	133	27.1
Sandy oxidised ware	13	5	1.1	46	0.8	10	2.0
Fine sandy grey ware	14	33	7.3	372	6.5	78	15.9
Malvernian metamorphic (H) (**MAL RE A**)	3	17	3.8	441	7.7	0	0.0
Malvernian metamorphic (W)	19	16	3.5	235	4.1	36	7.3
Total Regional ware		*391*	*86.3*	*4973*	*87.1*	*429*	*87.6*
Southeast Dorset BB1 (**DOR BB 1**)	22	24	5.3	225	3.9	29	5.9
Lower Nene Valley c. c. ware (**LNV CC**)	28	1	0.2	28	0.5	0	0.0
Oxfordshire red-slipped ware (**OXF RS**)	29	9	2.0	65	1.1	0	0.0
Mancetter Hartshill Mortaria (**MAH WH**)	32	6	1.3	146	2.6	10	2.0
Oxford red-slipped Mortaria (**OXF RS**)	33.3	1	0.2	26	0.5	9	1.8
Oxford white ware (**OXF WH**)	38	2	0.4	9	0.2	0	0.0
Oxford parchment ware (**OXF PA**)	40	2	0.4	4	0.1	0	0.0
Total Traded ware		*45*	*9.9*	*503*	*8.8*	*48*	*9.8*
Amphorae, Class 25 (**BAT AM 2**)	42.1	2	0.4	156	2.7	0	0.0
Samian	43	15	3.3	79	1.4	13	2.7
Total Imported ware		*17*	*3.8*	*235*	*4.1*	*13*	*2.7*
Total pottery		**453**		**5711**		**490**	

TABLE 4.2 SUMMARY OF THE ROMAN POTTERY ASSEMBLAGE BY FABRIC

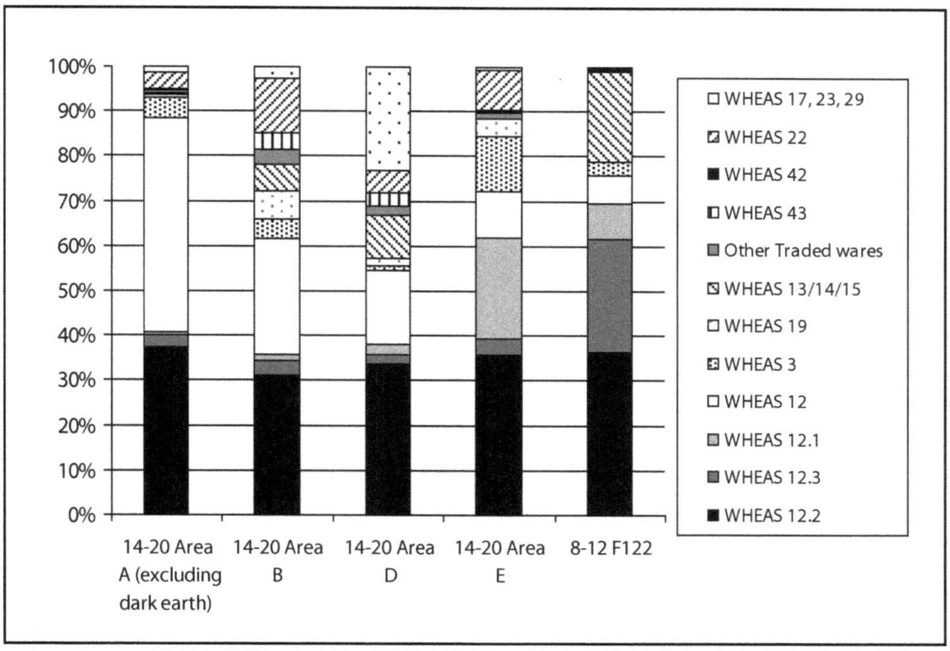

FIGURE 4.6 COMPARISON OF FABRICS FROM PIT 122 AND AREA GROUPS

period using the same methodology as for 14–24 The Butts. To avoid duplication, this is summarised only within the larger report (see Chapter 5), which also includes a list of all fabrics represented (Table 5.8).

POTTERY AND THE STRATIGRAPHIC SEQUENCE

Nineteen contexts produced Roman pottery. The largest groups came from a large pit (122), a ditch (104) and the 'dark earth' layers (110, 111, and 112) (Table 4.1). Only these groups are discussed in detail.

PIT 122

Of particular interest is the characteristically late 1st- to early 2nd-century, or 'Flavian-Trajanic', assemblage from pit 122. The pit was sealed by a gravel surface (117) which, as described below, also contained 1st- to early 2nd-century pottery, so the assemblage forms a well dated group. Pottery of a similar date was recovered from 14–24 The Butts (Fig. 4.6; Chapter 5). The earliest assemblages there were dated to the early 2nd century, based on the samian ware and associated Trajanic coins. These assemblages included rusticated jars, similar to those described below, and other typically early coarse ware forms.

Only one sherd of samian ware was recovered, not closely dated, and the only coin associated with pit 122 was Claudian, pre-dating the pottery. The coarseware forms were typically Flavian to Trajanic types (Fig. 4.7, 1–10). The most common fabrics by far were the characteristically early, organic-tempered Severn Valley ware variants (Table 4.2). These were more common here than in the early 2nd-century assemblage from Sidbury (Darlington and Evans 1992, fig. 5). They also seem more common than in the area assemblages from 14–24 The Butts, though this must partly reflect the presence of later pottery in these area assemblages. There was also a notably high proportion of reduced wares (Table 4.2. WHEAS 12.1, 12.3, 14); totalling 51% by weight compared to 6% in the overall assemblage from Sidbury. The reduced sandy ware (WHEAS 14) included a hard-fired fabric, similar to one noted at Alcester (Lee *et al* 1994, 13, fabric BC). The absence of all but a single sherd of BB1 supports a *terminus ante quem* of *c* AD 120.

Catalogue (Fig. 4.7)

WHEAS 12.2: Organic-tempered Severn valley ware, oxidised

1 FJ5.03. The assemblage also included the rim of a collared, Hofheim type flagon, similar to types noted in military contexts at Gloucester (Ireland 1983, fig. 67, 103) and Wroxeter (Timby *et al* 2000, fig. 4.49, F1.1). Diam. 100mm (12%). Context 120.

2 JN1.01. Narrow-mouthed jar with an out curving bead rim. The type is described by Webster as a predominantly mid 1st- to 2nd-century form (Webster 1976, fig. 1, A3). The rim has warped during firing and may be a waster. Diam. 14 0mm (12%). Context 121.

3 JW19.19. Wide-mouthed jar with a triangular rim, dated by Webster from the 2nd to 3rd century (*ibid*, fig. 4, C22). Diam. 290mm (7%). Context 119.

4 BA1.01/ TA1.01. Carinated bowl or upright walled tankard, with a bead rim, broken at the carination. Plain burnished. The bowls are dated by Webster to the 1st to 2nd century, and upright walled tankards to the mid to late 1st century (*ibid*, fig. 7, E38; fig. 9, H59, 60). Diam. 110m (22%). Context 121.

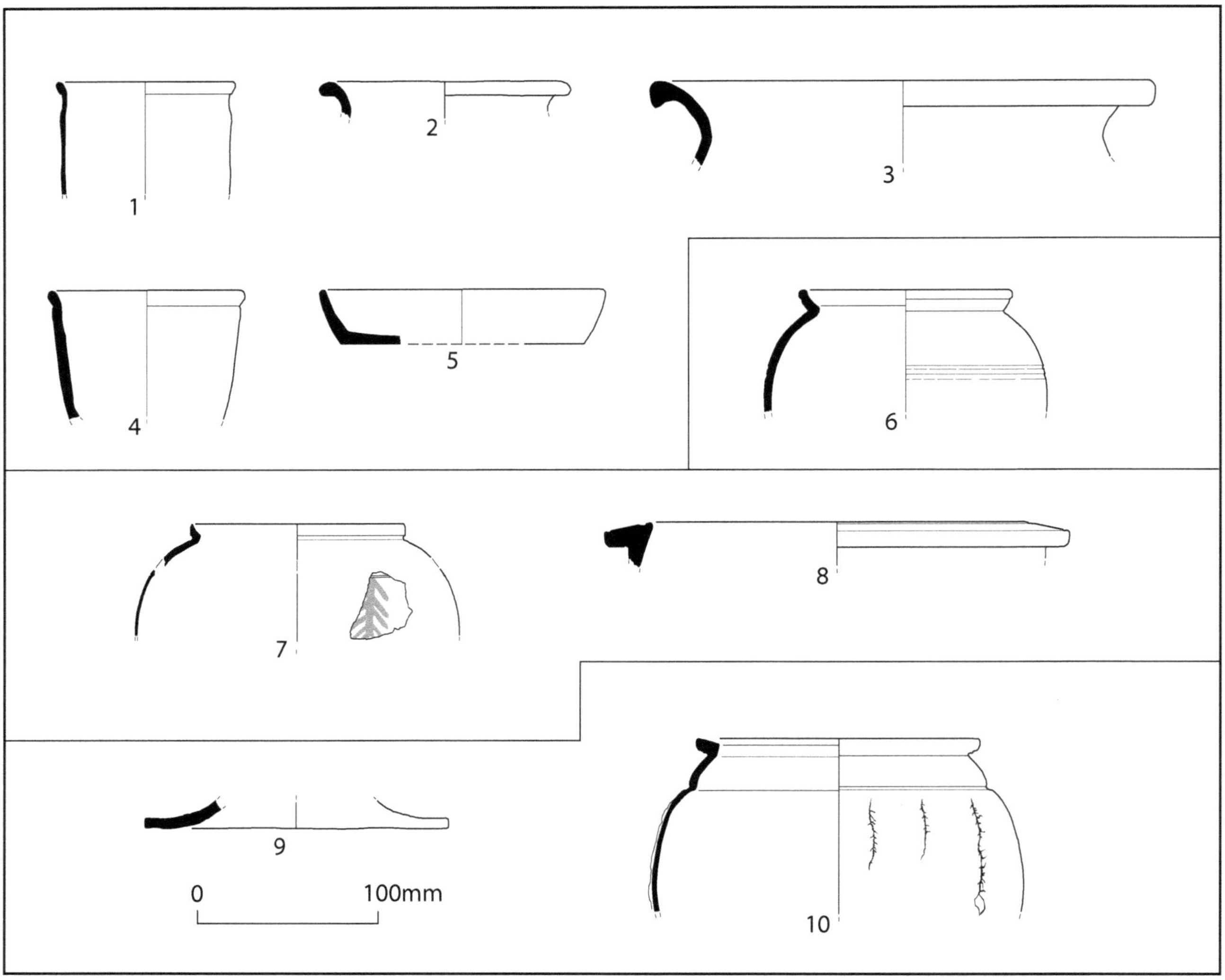

FIGURE 4.7 ROMAN POTTERY FROM PIT 122

5 DB16.01. Plain-rimmed, straight sided dish or platter; similar to examples found at Worcester, Deansway (Bryant and Evans 2004, fig. 157, 13–14) and Sidbury (Darlington and Evans 1992, fig. 20, 2–3). No exact parallels are published by Webster, but the Severn valley ware examples he illustrates generally date to the 1st to 2nd century (Webster 1976, fig. 10, K69-74). This would be supported by the use of an organic tempered fabric. Diam. 160mm (5%). Context 121.

WHEAS 12.3: Organic-tempered Severn valley ware, reduced

6 JM14.1. Globular beaker with an everted rim, which has a pronounced cup internally.

Burnished on the rim and shoulder, below which are two incised grooves, The body is decorated with linear rustication. Similar to forms noted at Worcester, Sidbury (Darlington and Evans 1992, fig. 22, 1) and Deansway (Bryant and Evans 2004, fig. 159, 5). Diam. 120mm (100%). Context (121). WHEAS 13: Sandy oxidised ware

7 BKD6.05. Globular beaker with an upright, internally cupped rim, thin walls and a delicate rim. The beaker is in a fine sandy fabric, similar to Alcester fabric DW (Lee *et al* 1994, 6), and is decorated with a white slipped herringbone pattern. The form is paralleled at Alcester (cf. *ibid*, fig. 28, O.194) and in the military assemblage at Wroxeter (Timby *et al* fig. 4.52, BK6). Diam. 50mm (10%). Context 120.

8 BI17.11. Bowl with a slightly down turned, grooved, flange rim, in a coarse sandy fabric. Similar to vessels found at Wroxeter, predominantly in 1st-century military contexts, but also in 2nd-century contexts (Timby *et al* 2000, fig. 4.67, B14.28/29). A similar vessel is also illustrated from Dodderhill, Droitwich (Rees 2006, fig. 20, 24). Diam. 210mm (4%). Context 121.

WHEAS 14: Fine sandy reduced ware

9 LAA15.08. Plain, angular rim from a concave lid; not datable. Possibly used in conjunction with the rusticated jar described below. Diam. 170m (28%). Context 123.

10 JGA7.13. Jar, with an everted rim and indented shoulder. Decorated with regularly spaced linear rustication, and burnished on the neck and rim. This form has been noted elsewhere in Worcester, for example at Deansway (Bryant and Evans 2004) and is paralleled at Alcester (Lee *et al* 1994, fig. 7, R135). Diam. 160mm (50%). Context 123.

GRAVEL SURFACES, CONTEXTS 117 AND 118

The pottery from the gravel surfaces sealing pit 122 (117) and the surface to the west of the pit (118) also dated to the late 1st or early 2nd century. Context 117 produced sherds of Severn Valley ware, predominantly organic tempered. Context 118 produced a characteristically early Severn Valley ware bowl type (cf. Webster 1976, fig. 9, H59) and five sherds of samian ware, possibly from Les Martres-de-Veyre, which exported samian between *c* AD 100 and 120 (Webster 1996, 3).

FEATURE 104

The pottery from the medieval ditch was more mixed, as would be expected from an assemblage that had been redeposited in a later feature. It contained some diagnostically early fabrics (WHEAS 12.2, 12.3, 3) and forms (cf. Webster 1976, fig. 4, C20), two sherds of South Gaulish samian ware, Black Burnished ware (BB1), a Mancetter-Hartshill hammerhead mortarium dating to AD 160+, and diagnostically later Roman fabrics (WHEAS 38) and forms (cf. *ibid*, fig. 7, E44).

THE DARK EARTHS

Three contexts represented spits excavated through a layer of 'dark earth' (contexts 110, 111, 112; Fig. 4.8). The pottery from these deposits, with an average sherd weight of 12g, was more fragmentary than the hollow way fill assemblages from 14–24 The Butts (see Chapter 5) and Castle Street (Dalwood *et al* 1997, 7, appendix 4, context 204). It seems less likely, therefore, that the assemblage from 8–12 The Butts derived from a midden, as argued for the two other assemblages. The material contained some of the characteristically late 3rd- to 4th-century fabrics (eg, WHEAS 29) and forms noted at 1 The Butts and 14–24 The Butts. Diagnostically late Severn Valley ware forms, represented by fragmentary rims, included a short-necked jar and splayed tankard (Webster 1976, fig. 7, C32/ 33, E44), and a pulley rim jar (*ibid*, fig. 3, A11). The groups also included residual sherds in early fabrics (WHEAS 12.2, 42, 43). The absence of the diagnostically late 4th-century + shelly wares is interesting, given their presence in the other late assemblages discussed in this volume.

The assemblage included sherds from a crudely formed vessel in handmade Malvernian ware (WHEAS 3). One rim was very similar to an example published from Worcester, Sidbury (Darlington and Evans 1992, fig. 35, 4c). The function of these vessels is uncertain; suggestions have included portable bread ovens and beehives (*ibid*, 67–69), but they could have had an industrial function. They are, however, characteristic of 3rd- and 4th-century deposits. A chunky, rather angular, rim had no apparent curvature and could be related (cf. *ibid*, fig. 35, 4a), although the

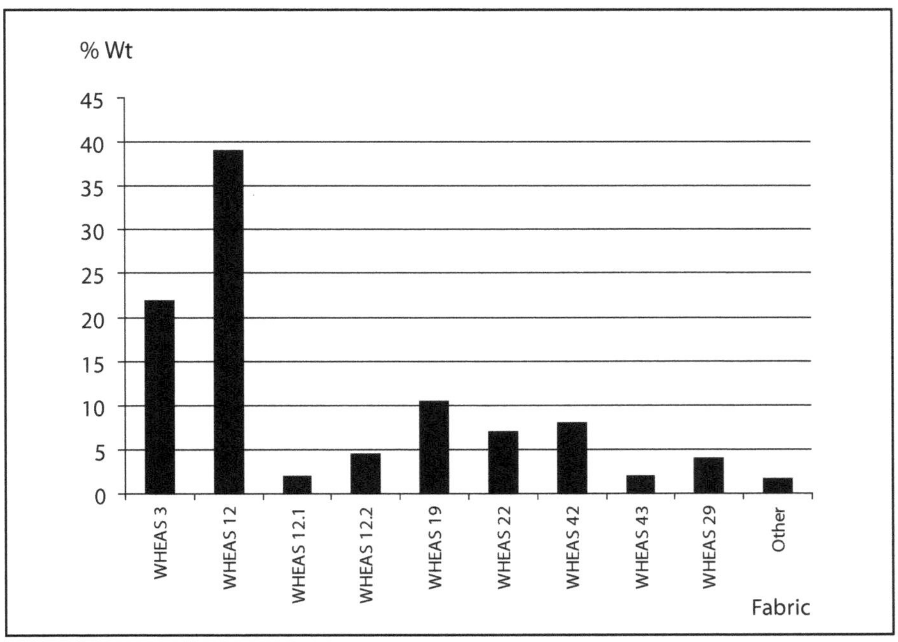

FIGURE 4.8 DARK EARTH FABRICS BY % WEIGHT

form is also similar to large storage jars found at a late Roman kiln site in Malvern (Peacock 1965–67, fig. 4, 81). Substantial quantities of oven material, some apparently *in situ*, have subsequently been seen by this author from recent excavations at the site of the new Library and History Centre Site in the Butts.

DISCUSSION

This small evaluation produced a well-dated early Roman assemblage from a sealed deposit (122). The presence of forms that, elsewhere, are associated with military activity is worth noting, perhaps adding to the growing body of artefactual evidence for some form of military presence in early Roman Worcester. The assemblage from the dark earth deposit contributes to the debate about what these dark earths represent in Worcester, discussed in more detail later in Chapter 5 of this volume.

ENVIRONMENTAL EVIDENCE FROM PIT 122

Elizabeth Pearson

The sample was processed by flotation followed by wet sieving using a Siraf tank. The flot was collected on a 300 micron sieve and the residue retained on a 1mm mesh which allows for the recovery of items such as small animal bones, molluscs and seeds. The residue was fully sorted by eye and the abundance of each category of environmental remains estimated. The flot was fully sorted using a low power EMT stereo light microscope and plant remains identified using modem reference collections maintained by the Service, and seed identification manual (Beijerinck 1947). Nomenclature for the plant remains follows the *Flora of the British Isles*, 3rd edition (Clapham *et al* 1989). Animal bone was identified with the aid of modem bone reference collections housed at the Historic Environment and Archaeology Service and identification guides (Schmid 1972; Hillson 1992).

A total of 3.4 kg (over 95 fragments) of large mammal bone were recovered from the 201 sample from context 123. The results are listed in Table 4.3. These were predominantly well-preserved, mostly entire horse (*Equus*) bones which are likely to have come from one individual as pairs of bones were recorded. No butchery marks or pathology were noted on these bones, and no definite signs of unfused epiphyses. It is therefore likely that this is a disarticulated skeleton of a mature animal which has not been butchered. The bones are likely to have been rapidly buried and not exposed on the surface for long, or redeposited from elsewhere on the site, as they are well preserved, little fragmented, and show no signs of gnawing by dogs or

Species	Element	Unfused	Measurable	No	Comments
Bovis	ulna		0	2	
Equus	patella		y	1	
Equus	humerus			1	Part of shaft and prox end missing
Equus	femur		y	1	
Equus	humerus		y	1	Likely pair of humerus above
Equus	metatarsal		y	2	
Equus	radius+ulna		y	1	Radius whole, ulna distal
Equus	pelvis		0	1	
Equus	metacarpal		0	2	pair
Equus	radius		0	1	
Equus	phalange		0	1	
Equus	tarsal		0	2	pair
Equus	calcaneus		0	1	
Equus	vertebrae		0	2	thoracic
Equus	tibia		0	1	
Equus	tarsal		0	1	
Equus	radius	?	0	1	Distal end
Equus	pelvis		0	1	
Ungulate	femur		0	2	
Ungulate	rib		0	1	
Ungulate	vertebrae		0	1	
Ovis/ capra	molar		0	1	
Ovis/ capra	horncore		y	1	
Ovis/ capra	horncore		0	1	
Ovis/ capra	tibia		y	1	
Ovis/ capra	ulna		0	7	
Sus	molar		0	1	
Ungulate	unidentified		0	50–60	
Ungulate	rib		0	5	

TABLE 4.3 ANIMAL BONE FROM SOIL SAMPLE OF CONTEXT 123

Latin name	Family	Common name	Habitat	No
Hordeum vulgare grain	gramineae	barley	cultivar	2
Cereal sp idet grain	gramineae	cereal	cultivar	1
Gramineae sp idet grain (large)	gramineae	grass	cultivated ground/ cultivar	2
Gramineae sp idet grain (small)	gramineae	grass	cultivated ground/ cultivar	5
Montia Fontana ssp chondrosperma	portulaceae	blinks	Aquatic/ wet habitat	1
Chenopodium album	Chenopodiaceae	Fat hen	Cultivated/ disturbed ground	1
Chenopodium glaucum/ rubrum	Chenopodiaceae	Glaucus/ red goosefoot	Cultivated/ disturbed ground	1
Atriplex sp	Chenopodiaceae	orache	Cultivated/ disturbed ground	1
Vicia/ Lathyrus sp (small)	Leguminosae	Vetch/ vetchling/ pea	Cultivated ground	4
Melilotus/ medicago/ trifolium sp	Leguminosae	Melitot/ medick/ clover	Cultivated/ disturbed ground	2
Rumex acetosella agg	Polygonaceae	Sheep's sorrel	Cultivated/ disturbed ground, grassland meadows and heathland	5
Cf Rumex acetosella agg	Polygonaceae	Sheep's sorrel	Cultivated/ disturbed ground, grassland meadows and heathland	1
Urtica dioca	Urticaceae	Common nettle	Woodlands hedgerows scrub grassland meadow and heathland	1
Eleocharis sp	Cyperacae	Spike-rush	Aquatic / wet habitats	4
Carex sp	Cyperacae	sedge	Woodlands hedgerows scrub grassland meadow and heathland. Aquatic/ wet habitats	1

TABLE 4.4 PLANT REMAINS FROM SOIL SAMPLE OF CONTEXT 123

rodents. A small number of remaining bone fragments included sheep/ goat (*Ovis/ Capra*) and pig (*Sus*) molars and sheep or goat long bones and horn core.

The soil sample was charcoal rich and contained a small assemblage of charred plant remains. These are listed in Table 4.4. The assemblage is dominated by small weed seeds which may have been growing with a cereal crop as occasional cereal grains (barley; *Hordeum vulgare* and unidentified cereal) were recovered. Cereal crop waste may have been used as tinder or fuel for fires and disposed of in this pit, but equally the burnt seeds may simply have been present in other waste. The burnt waste and the animal bone seem to be unrelated as there were no signs of burning on the bones.

DISCUSSION

The deposits observed in the evaluation represent an unusually well-preserved sequence of Roman deposits, apparently sealed in part by later Roman or immediately post-Roman 'dark-earth' soils. The earliest Roman material (late 1st to early 2nd century) appears to pre-date industrialisation of the Roman settlement, but fragments of ceramic building material, probably *lydion* (a long flat brick), clearly indicate the presence of masonry buildings in the vicinity at this early stage. The early pit (122) was sealed by a gravel surface (117), over which numerous large and unabraded sherds of late 1st- to early 2nd-century Severn Valley ware were scattered. Whilst it would be tempting to interpret this as an early gravelled road surface (as seen by Barker to the south), the observed evidence is rather against this interpretation. The gravel layer was thin and not obviously worn or compacted, and sherds deposited on a roadway would very soon become crushed and abraded, which was demonstrably not the case here. The surfacing possibly represents some roadside yard, but was clearly not exposed to great levels of foot or wheeled traffic. The small area of pebbled surfacing (118) observed in the central part of Trench 1 was even thinner, and consisted of finer grade stones; possibly it was an internal surface. It was, however, very compact and the pottery crushed into the surface showed it had experienced some wear. The surfacing was not, however, in any way comparable with the previously observed road metalling at Broad Street (Barker 1968–69b, 63–64) which had built up to in excess of 1m thick. In the observed section on the present site, the slag layer was no more than 0.35m deep. The dating of the slag surface here (to the 2nd century) also appears somewhat earlier than the 3rd- to 4th-century dating of the Blackfriars road.

It is of course feasible that the later Roman road metalling was deliberately removed from this area, possibly by the iron master Andrew Yarranton in the late 17th century, but this seems unlikely, given the extreme difficulty of removing concreted slag which has both been compacted and fused by rust. Iron-rich slag could be freely recovered from dumps close to the surface in this area, and was readily available from riverside deposits at Cinder Point. The sequence of post-medieval deposits within the evaluation trenches also does not indicate organised quarrying to the road levels, but only piecemeal pit digging into the sand and gravel beneath.

The apparent absence of the road from the eastern part of the site might suggest that it passed further to the west than previously thought (the western end of Trench 1 had been severely affected by post-medieval pit digging), although work by Birmingham Archaeology on the neighbouring site at 14–24 The Butts found no trace of the road (see Chapter 5). A course to the east (under the Paul Pry Public House) is feasible, but would require a substantial curve in the postulated alignment.

Later Roman or post-Roman activity on the site is attested by the presence of stake-built structures post-dating an accumulation of 'dark-earth' material overlying Roman gravel and slag surfacing. The apparent absence of contemporary artefacts gives the initial impression that these may have been animal pens or enclosures, but only further excavation of the site may clarify the nature of these structures.

The presence of an apparently near full sequence of Roman deposits immediately beneath modern ground levels indicates the vulnerability of such significant deposits to even minor groundworks. Here the preservation to high level may be due to burial by Civil War period earthworks, as the best preserved area was immediately behind the probably 17th-century ditch line.

Small but significant quantities of medieval tile wasters were recovered from this site, indicating some tile making activity in the near vicinity. This is not unexpected given the extra-mural location close to Blackfriars monastery, and the presence of both sand and clay in the immediate area.

The post-medieval features were not studied in detail, but sufficient of ditch 104 was excavated to suggest that it was of mid to late 17th-century origin and was possibly defensive. The documented existence of a Civil War earthwork in this area strengthens the evidence for ditch 104 originating as part of an earthwork bastion.

CONCLUSION

The site produced evidence of early Roman activity in this area, which appears to have included masonry buildings at an early stage in the formation of the settlement. The building materials here may be related to the evidence for a masonry building seen at 1 The Butts, and in observations at Tramp's nightclub to the south. Little evidence of masonry structures has been seen in the area of 14–24 The Butts to the west, nor during previous works at Farrier Street to the northeast.

Very little of the Roman archaeology of the site was exposed in plan during the evaluation, and it was therefore difficult to characterise the activity. The original function of pit 122 may have related to the extraction of sand for constructional purposes. The subsequent sequence of surfaces may relate to yards, or possibly a roadside market function – the abundance of unabraded pottery on surface 117 supports either hypothesis.

Post-Roman activity appears to have been primarily agricultural, with apparently small scale sand extraction in the post-medieval period. There was little evidence for medieval activity, and the artefacts recovered may well reflect manuring and deposition of waste in sand pits. It is, however, possible that any more substantial medieval deposits were truncated by the creation of Civil War earthworks. The location and limited scale of the evaluation trenches has not provided a clear characterisation of the Civil War defences; it seems likely, however, that the cut elements survive. A probable Civil War defensive ditch has also been recorded at 14–24 The Butts (see Chapter 5) but it is unclear how the two may relate.

The evaluation has identified the presence of well-preserved Roman deposits surviving to a high level in the northeastern half of the site. The full extent of these deposits is presently unclear, but there appears to be potential for further deposits or cut features of this date across the remainder of the site, except where locally truncated by cellarage and deeper post-medieval features. Possibly post-Roman 'dark-earths' also survive in the northeastern corner of the site, immediately beneath the floors of the recently demolished buildings. Redevelopment of the site in 2008–10 was designed with a view to avoiding any further impact on the archaeological remains and to ensure their preservation.

There appears to be the potential to recover something of the plan of the Civil War defences in this area, but the complex of intercutting extraction features at the western end of the site may have removed crucial evidence. Overall, the site appears to have significant archaeological potential, and may prove to be the key in linking previous observations at Farrier Street, 1 The Butts and 14–24 The Butts.

ACKNOWLEDGEMENTS

Archaeological fieldwork was undertaken by Mike Napthan Archaeology between September and October 2003 prior to a residential development. The assistance and cooperation of the following was much appreciated: the architect Roger Dean Walker of Dean-Walker Bateman and the client Mr Andrew Grant. Maxine Chong and

Kerry Whitehouse assisted with the fieldwork and post-excavation processing. Specialist reports were contributed by C Jane Evans and Liz Pearson. The project was monitored on behalf of Worcester City Council by James Dinn (Archaeology Officer). Copy editing was carried out by Malcolm Hislop.

CHAPTER 5: EXCAVATIONS AT 14–24 THE BUTTS

Richard Cuttler, Robert Burrows, and C Jane Evans

with

Robert Bracken, Nick Garland, Emma Hancox, Isabel Nogueira, Erica Macey-Bracken, Helen Martin-Bacon, Jon Millward, Stephanie Rátkai, Sian Reynolds, David Smith and Wendy Smith, Emma Tetlow, Jane Timby, Roger Tomlin, David Williams, and Felicity Wild

SUMMARY

The earliest remains at 14–24 The Butts were dated to the 2nd and early 3rd centuries (Phase 1). These were mostly in the form of shallow pits and ditches, thought to be the result of small-scale gravel quarrying. The northern extent of this activity appears to have been defined by a large east–west aligned ditch, with outlying stakeholes, which may have formed a boundary ditch and fence. No features associated with ironworking were recorded, although the ubiquitous presence of iron waste products dating from both Phase 1 (2nd and early 3rd centuries) and from Phase 2 features (3rd and 4th centuries) indicate significant Roman period iron smelting nearby.

Further gravel quarrying during the 4th century (Phase 2) was contemporary with a cobbled surface, interpreted as a Roman thoroughfare, which replaced the earlier boundary fence. This thoroughfare formed within an east–west hollow way, which possibly remained in use beyond the Roman period and was subject to frequent repair. It was sealed by a garden-like soil which filled the hollow way, and which contained significant quantities of residual Roman finds, but also medieval and post-medieval material. Despite the presence of 4th-century (Phase 2) material, the medieval and post-medieval (Phase 3) finds demonstrate that this feature was either backfilled or that it was intruded upon in (possibly by ploughing) in the late medieval or post-medieval period.

The site revealed very little stratigraphic or artefactual evidence for activity between the late Roman and late medieval periods. A large ditch, aligned east–west across the southern extent of the site, contained some medieval pottery and tile, and may relate to defensive activity during the Civil War. Archaeological features in other parts of the site were heavily truncated with some areas totally destroyed by post-medieval activity, which may also explain the absence of features dating from between the late Roman and late medieval periods.

In addition to the excavation, the laying of ring beam foundations and the removal of contaminated ground required an archaeological watching brief to the deepest level of ground disturbance. Further pits, postholes and the Roman cobbled surface were recorded in plan, but not excavated as they were below the depth of the development. The watching brief also recorded a stone-lined well, the fill of which contained waterlogged plant and insect remains from the Roman period. These suggest that during this period the site may have been close to pasture, or lay on the margins of habitation.

INTRODUCTION

The development area comprised approximately 0.45ha (NGR SO 8473 5512) and is located on the north side of The Butts (Figs 1.3 and 5.1). The site is entered from the east by Netherton Lane, and there is a council yard to the west and a railway viaduct to the north. Early maps depict the site as largely open ground until the appearance of Netherton House on Doharty's map of 1741. During the 19th century, the site continued to be occupied with buildings associated with Netherton House, with further buildings added around the turn of the 20th century. Two garages are depicted on the 1965 Ordnance Survey map and by 1970 an electricity substation had been constructed on the site.

Proposals for the construction of 77 residential units with associated car parking prompted an archaeological evaluation in July 2000. By this time, the garages had been demolished and the site was used as an NCP car park. Four trial trenches (Fig. 5.1, Trenches 1–4) demonstrated the presence of significant numbers of discrete features cutting into the natural subsoil within the eastern half of the site (Trenches 1–3). These features comprised both Roman and post-medieval deposits, with little or no evidence for activity in the intervening period. Thus, approximately 1m of post-medieval overburden directly sealed the Roman deposits, giving the general impression that any intervening deposits from the late Roman and medieval periods had been substantially truncated in the post-medieval period.

The westernmost trench (Trench 4) produced several Roman artefacts, but these were found in association with finds of post-medieval date (glass, tile and pipe stem), and it was considered that the Roman material in this trench was probably redeposited rather than being in a primary context. The archaeological potential of the site therefore seemed to lie predominantly in the eastern half. This disparity between the eastern and western halves of the site may be due to levelling during the 19th and 20th centuries.

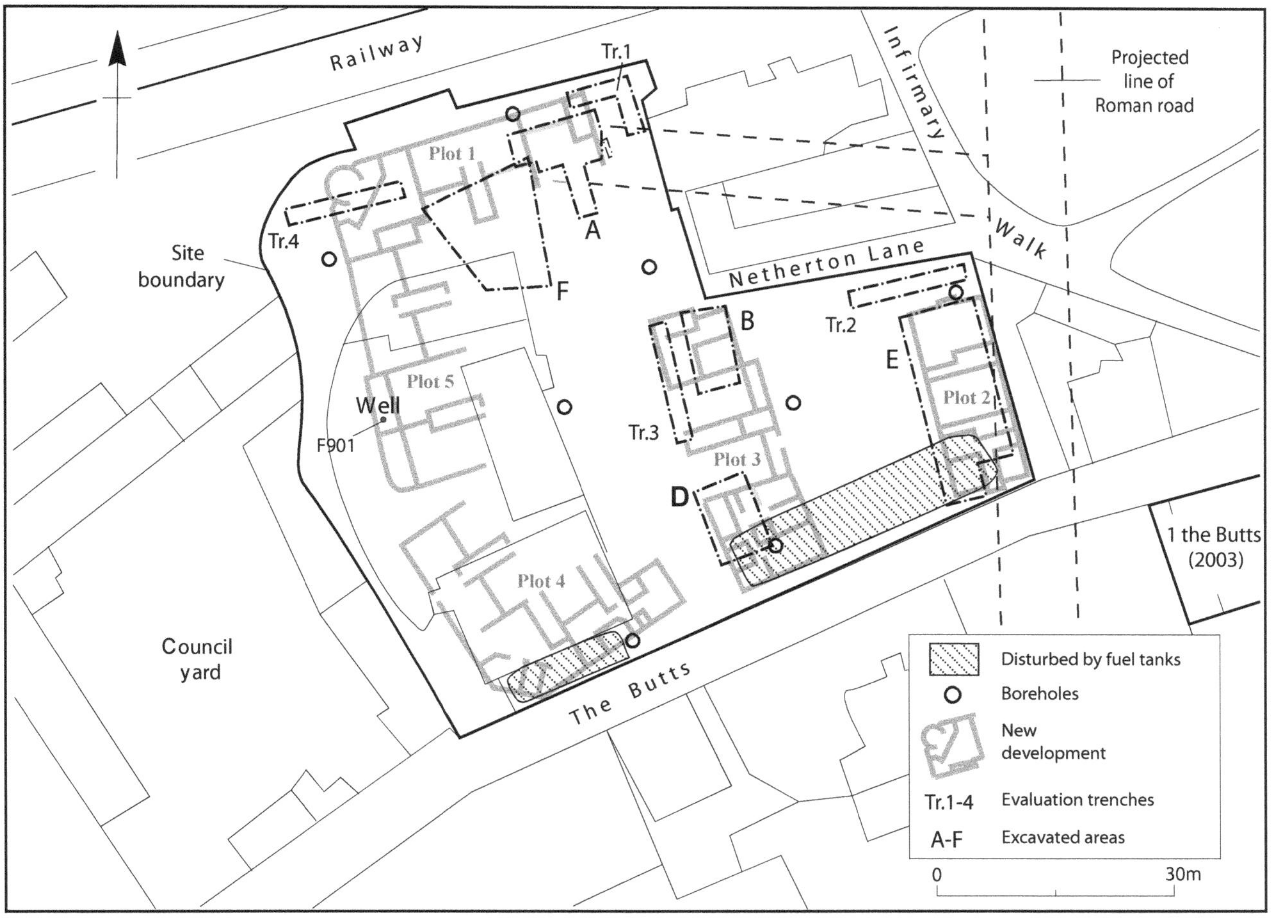

FIGURE 5.1 GENERAL SITE PLAN

Based on the results of the evaluation, four areas (A, B, D and E, Fig. 5.1) within the eastern half of the site were subject to further investigation, all targeting areas of major impact. Areas A, B and D measured approximately 10m by 7m, while Area E measured 24m by 10m. Two of the areas (A and B) were designed to examine features found during the evaluation (Trenches 1 and 3), and one area (Area D) was designed to examine the northern side of The Butts's street frontage. This was considered of particular interest given that its archaeological potential had not been evaluated during previous investigations which had focused on the southern side of the street (Bretherton and Pearson 1998). Area E was designed to investigate the projected north–south alignment of the former Roman road discovered during excavations at Broad Street and Blackfriars to the south (Barker 1968–69). The greater depth of overburden in the southwestern part of the site suggested that any archaeological deposits here were not at risk and that mitigation was unnecessary.

Modern overburden was removed using a 360 degree tracked excavator to the uppermost surface of the archaeological deposits. Further sampling was by hand excavation. Once building works commenced the site was subject to continual archaeological monitoring during ground works and the removal of contaminated ground in the northern half of the development. An area (Area F, Fig. 5.1), of approximately 100m^2, was stripped of overburden to reveal the natural subsoil at a depth of approximately 0.3–0.4m. Where features were to be preserved *in situ,* they were photographed and planned. The groundworks for ring beams involved the opening of trenches around concrete piles where archaeological deposits were hand-excavated prior to the lowering of the ground level. As a similar numbering sequence was used for both the evaluation and excavation, features and contexts recorded during the evaluation have been given the suffix E.

RESULTS

The degree of truncation and depth of overburden varied significantly across the site. Within Area A the overburden was generally around 1m in depth, whereas further to the south (Areas B, D, E and Trenches 2 and 3), the overburden ranged between 0.51m and 1.35m. The deepest overburden was recorded at the western extent of the site (Trench 4) where it reached 1.75m in depth. No prehistoric deposits were recorded, although some features were devoid of artefacts, leading to the suspicion that they may be slightly earlier, but none conclusively dated earlier than the 2nd century (Phase 1). Some features recorded as part of the watching brief were not excavated (preserved *in situ*), and therefore have no associated finds. For this reason these features have been excluded from the phased results, but are described at the end of this section.

On the basis of stratigraphic relationships and finds, features can be placed into four phases of activity:

Phase 1: 2nd and early 3rd centuries
Phase 2: 4th century
Phase 3: 17th and 18th centuries
Phase 4: 19th and 20th centuries

PHASE 1: 2ND AND EARLY 3RD CENTURIES

AREA A AND TRENCH 1 (Figs 5.2–5.3, Plate 5.1)

In Area A, the natural subsoil (1030) was cut by three possible stakeholes (F110–F112), each measuring between 0.07–0.09m in diameter and 0.06–0.08m deep. None of these produced any finds, and were all sealed by light brown sandy silt (1034) varying between 0.24m and 0.35m in depth, which did produce a few Roman sherds. This appeared to be similar to a 'disturbed natural' layer at the southern extent of Area A (1014, 1035, and 1036). Layer 1034 was in turn cut by an oval pit (F113) towards the northern extent of Area A, measuring 0.60m in diameter, 0.38m deep with near vertical sides and a flat base.

At the southern extent of Area A was an east–west aligned ditch (F107/ F109), measuring approximately 3.70m wide, with a U-shaped profile. The ditch was 0.90m deep and had three distinctive fills. While the primary fill (1033, 1032) was devoid of finds, a second fill (1023, 1028) produced Roman sherds broadly dating to the 2nd century. The upper fill (1020, 1015) contained over 200 sherds of Roman pottery, including a Central Gaulish samian ware beaker rim (2nd century), together with animal bone and daub. Along the northern extent of the ditch was what may have been a later cut or the result of slumping (1017). This was further cut by a small pit (F106, 1016), 0.24m deep, which contained an almost complete 2nd-century *tazza* (Evans, this chapter) and a sherd of samian ware dated to the Antonine period (Wild, this chapter). The upper

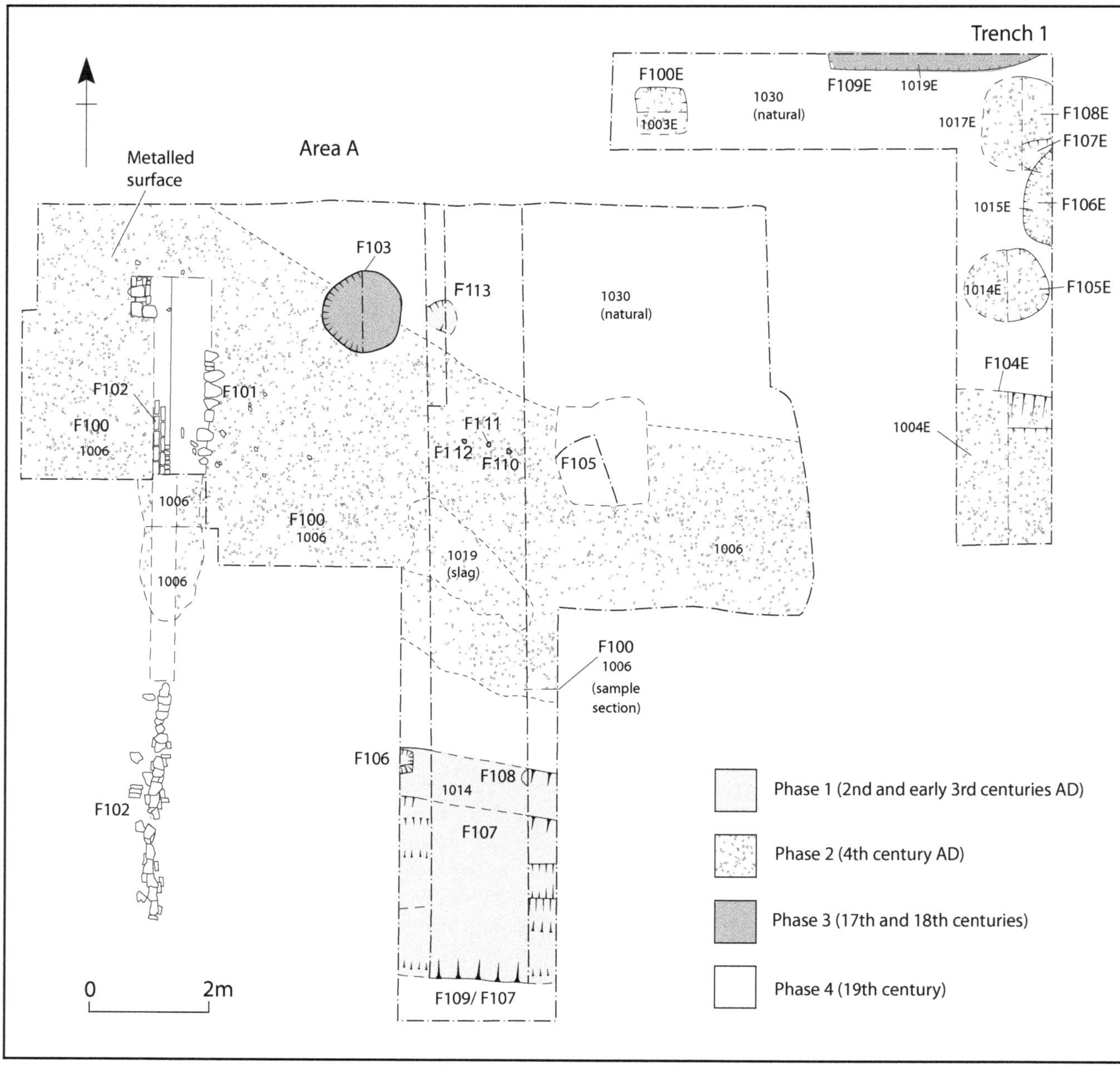

FIGURE 5.2 PLAN OF AREA A AND TRENCH 1

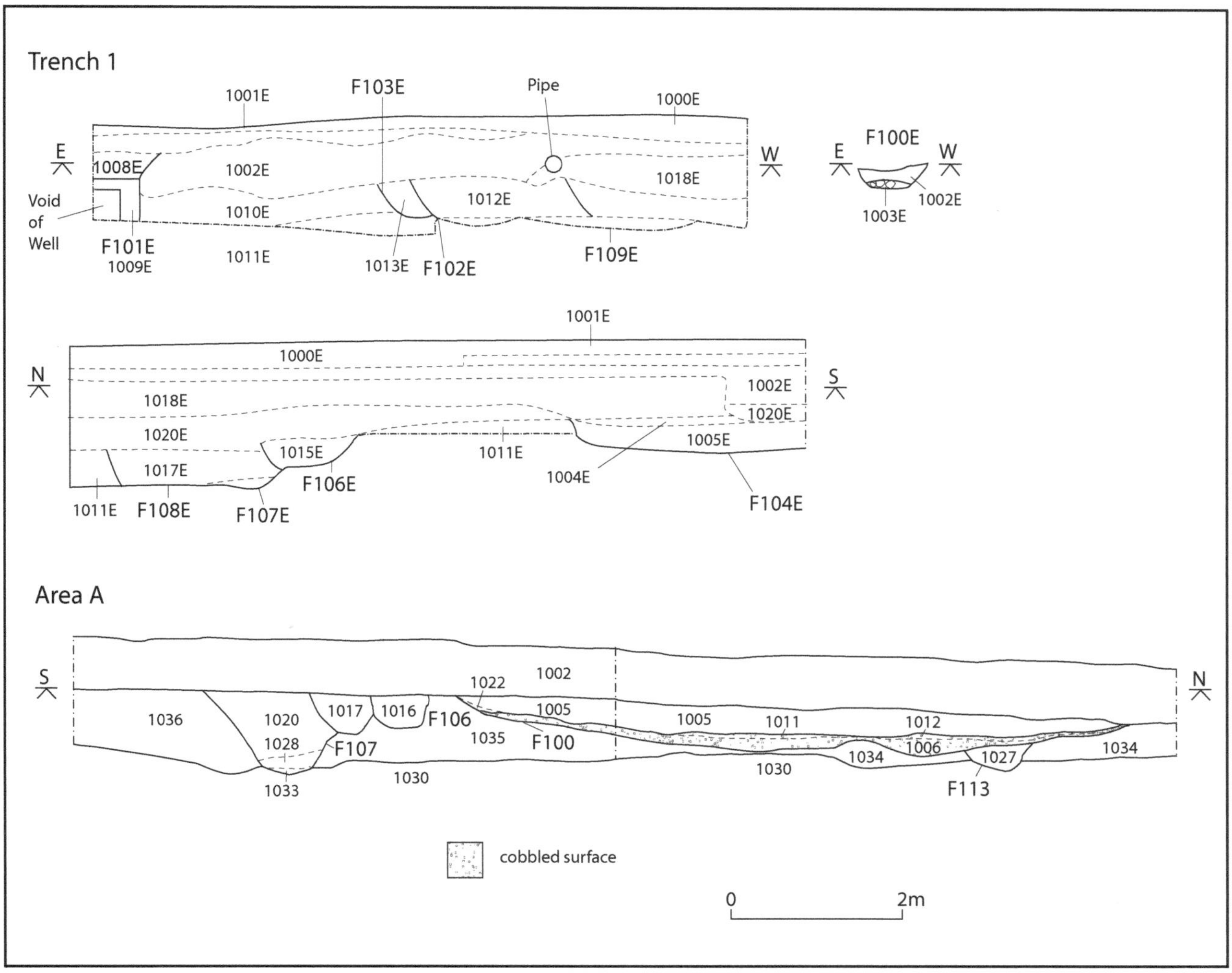

Figure 5.3 Sections from Area A and Trench 1

Plate 5.1 Phase 1 stakehole features (F110–F112) in Area A

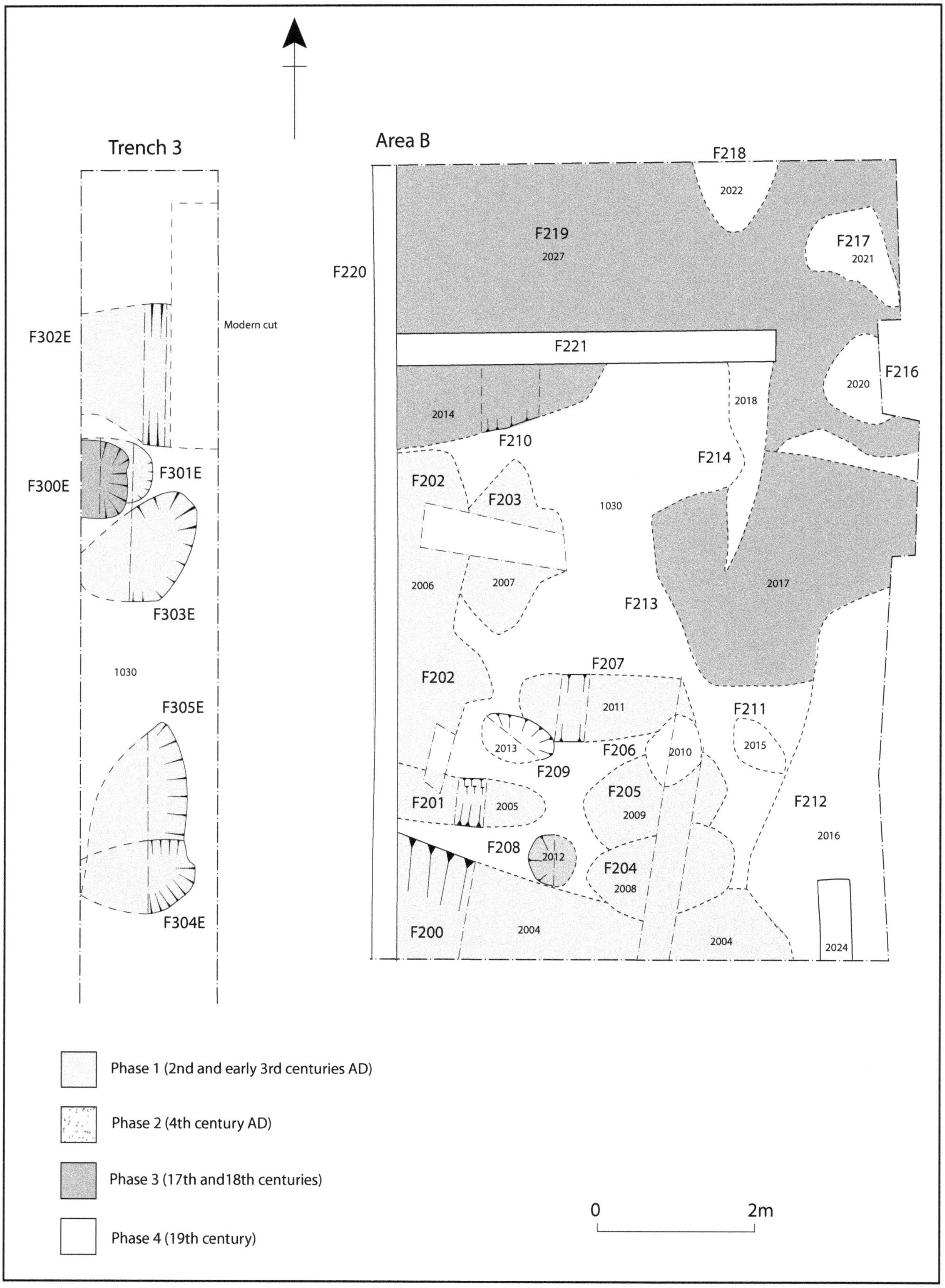

FIGURE 5.4 PLAN OF AREA B AND TRENCH 3

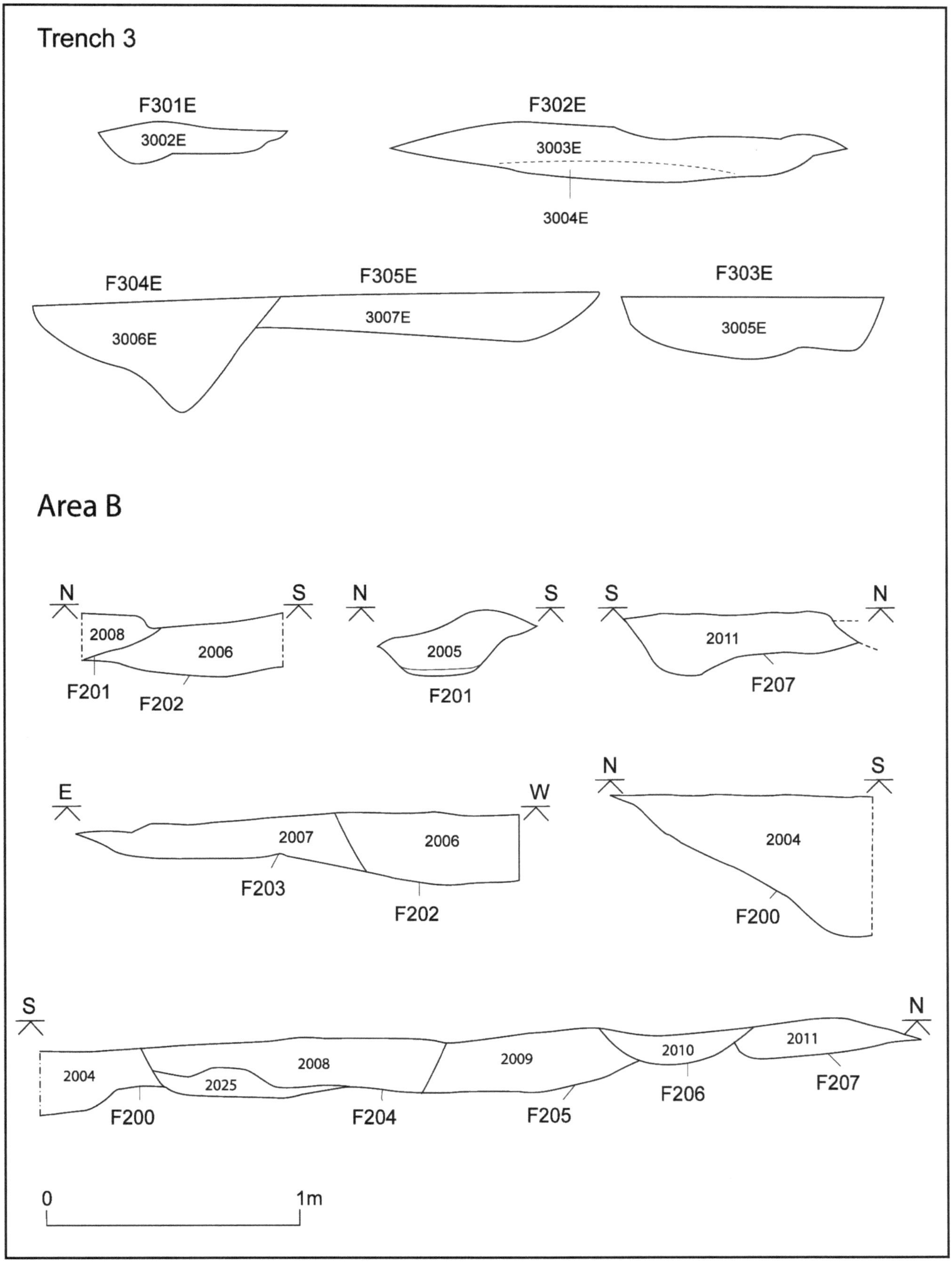

FIGURE 5.5 SECTIONS FROM AREA B AND TRENCH 3

PLATE 5.2 PHASE 1 PITS AND GULLIES IN AREA B

ditch (F107/ F109) fills were also cut by a second small pit (F108), which contained a moderate amount of 2nd-century Roman pottery, tile, iron slag, animal bone and window glass (1018).

AREA B AND TRENCH 3 (Figs 5.4–5.5, Plate 5.2)

Most of the features in Area B comprised pits and short gullies. The earliest feature is probably a small gully (F207) which produced Roman pottery, iron slag, animal bone and a very well preserved coin of Trajan (*Dupondius*, AD 101–2).

The remainder of the pits within Area B (F200, F202, F203, F204, F205 and F208) can be broadly dated to the 2nd and 3rd centuries. Pits F200 and F205 had been cut by an oval-shaped pit (F204). These pits were typically filled with medium grey-brown silty sand and contained Roman pottery, iron slag and animal bone. The pits ranged in size from 0.65–2.05m in diameter and generally 0.40m in depth. An irregularly-shaped pit (F203) had been cut by a possible large, also irregularly-shaped pit (F202), which, in turn, was cut by an east–west aligned butt ended linear gully (F201). While the fill of F201 (2005) produced very little in the way of finds, the gully clearly cannot be earlier than F202, which contained 2nd- to 3rd-century pottery.

At the southern limit of excavation was the northern edge of a large pit (F200). This measured in excess of 1m in depth and the fill (2004) produced 191 sherds of Roman pottery, including four sherds of Antonine samian ware. Also recovered were fragments of tile, animal bone, iron slag, nails and a *Denarius* of Septimius Severus or family (AD 193–211).

In Trench 3, four very heavily truncated pits and one ditch (F301E–F305E) all produced pottery forms that can be broadly dated to the 2nd or early 3rd century. One shallow pit (F305E), in the middle of the trench, was cut by a second pit (F304E) with a V-shaped profile. F303E was an irregular shaped, round-bottomed pit, 0.20m in depth and filled with a dark brown clay-silt-sand (3005E). F301E could have been the remains of a rounded ditch terminal or an irregular pit. At the northern extent of the trench, F302E, which may be the remains of a heavily truncated, U-shaped ditch, produced sherds of 2nd-century date from the lower fill (3004E). One further bowl-shaped pit (F301E) had been heavily truncated by a later 18th-century pit (F300E).

AREA D (Fig. 5.6)

Most of the features within Area D produced finds broadly dating to the late 2nd century. A layer of cobbles (3001) may have formed a surface (F300) on a roughly east–west alignment. The surface was 0.25m in depth and extended beyond the eastern and western limits of the area. This produced samian ware dated Hadrianic to Antonine. Other finds included tile, animal bone, iron slag and nails. F300 was cut by a small circular pit (F302), which was devoid of finds.

Located towards the southern edge of the area was a Roman pit (F305) measuring approximately 1.20m across. Second-century pottery, iron slag and four fragments of *tegula* were recovered from the fill (3006). The pit had been truncated by what appeared to be an east–west aligned ditch (F304) which terminated within the excavation area. The ditch measured in excess of approximately 3.40m in

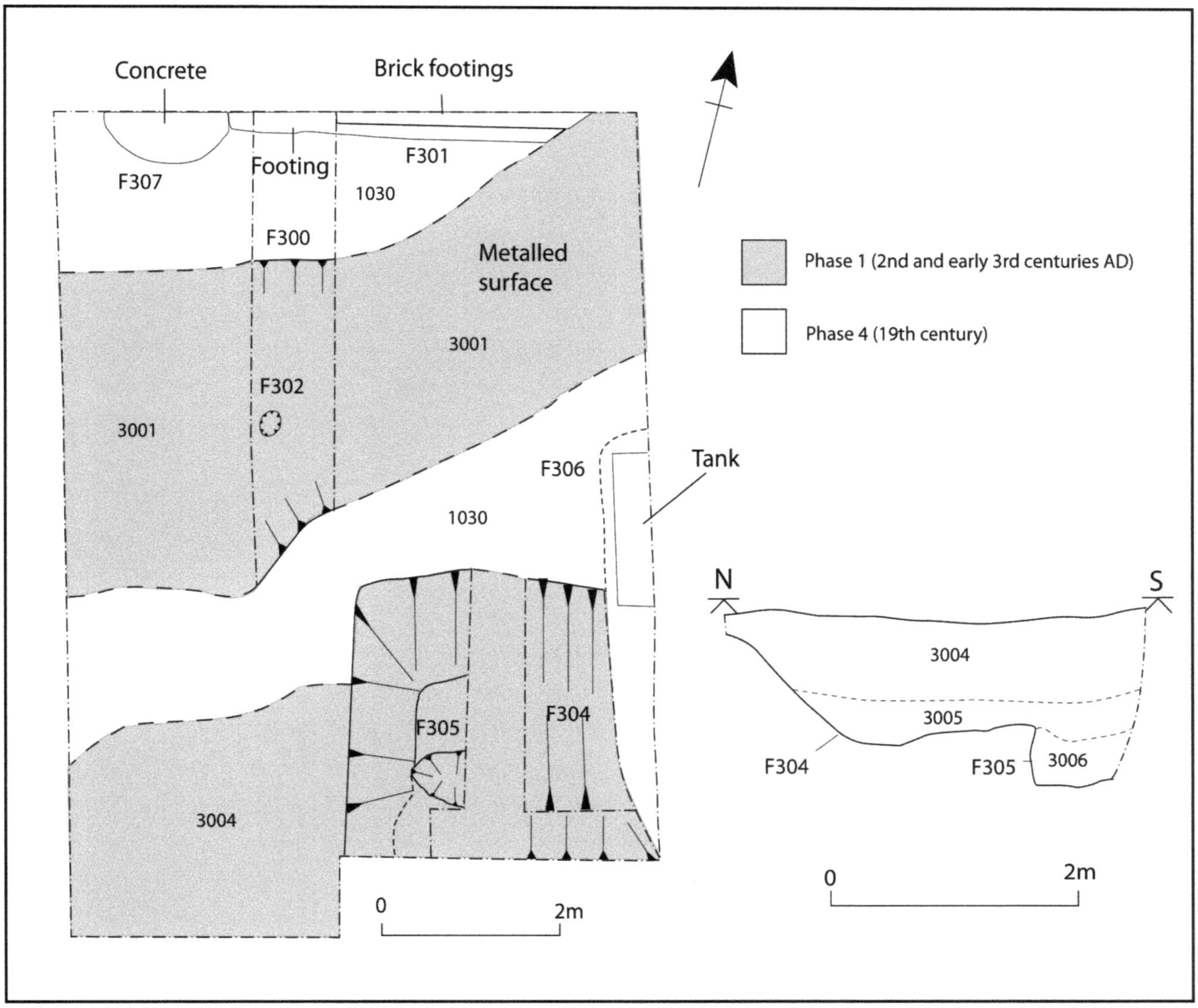

FIGURE 5.6 PLAN OF AND SECTION FROM AREA D

width, and the northern edge sloped sharply to a rounded base. A large assemblage of late 2nd-century pottery was recovered from the lower and upper fills (3005 and 3004 respectively). This included five sherds of samian ware dated to the Hadrianic period and mortaria; brick (notably *tegula*), iron slag, animal bone, glass and daub were also recovered.

AREA E AND TRENCH 2 (Figs 5.7–5.8)

Figure 5.7 shows that Roman remains only survived at the northern extent of Area E, and within Trench 2, where several pits (F200E, F402, F403, F411 and F412) were cut into the natural orange sand (1030). Some of these were fairly small (F402, F403, F411 and F412), measuring between 0.35 and 0.60m in diameter and 0.35m in depth. Pits F402, F411 and F412 produced iron slag, and one pit (F402, 4007) produced a 2nd- to 3rd-century tankard fragment.

In the northeast corner of Area E was what appeared to be the terminal of a ditch (F405). Both the upper (4013) and lower fills (4027) contained 2nd-century Roman sherds, including an almost complete Malvernian ware vessel. The assemblage also contained most of a large rusticated pot, amphora and the base of a samian ware vessel with the stamp of Lottius, dated AD 125–50. The fills and orientation of this ditch suggests that it is the same as a ditch in Trench 2 (F201E) which also produced pottery broadly dating to the 2nd century.

Ditch F405 was truncated to the west by a rubbish pit (F404), excavated to a depth of 1.50m. Both the upper and middle and upper fills (4015 and 4012 respectively) contained iron slag, well-preserved animal bone and pottery broadly dated to the 2nd century.

At the eastern end of Trench 2 was a possible pit (F200E) surviving to a depth of 0.4m. A possible wire earring was found in the fill (2003E), and pottery broadly dating to the 2nd century. It seems likely that this feature defines the northwestern edge of F420.

INTERPRETATION OF PHASE 1

Pottery dating to the 1st century was recovered from Phase 1, but only from features which also contained later, mostly

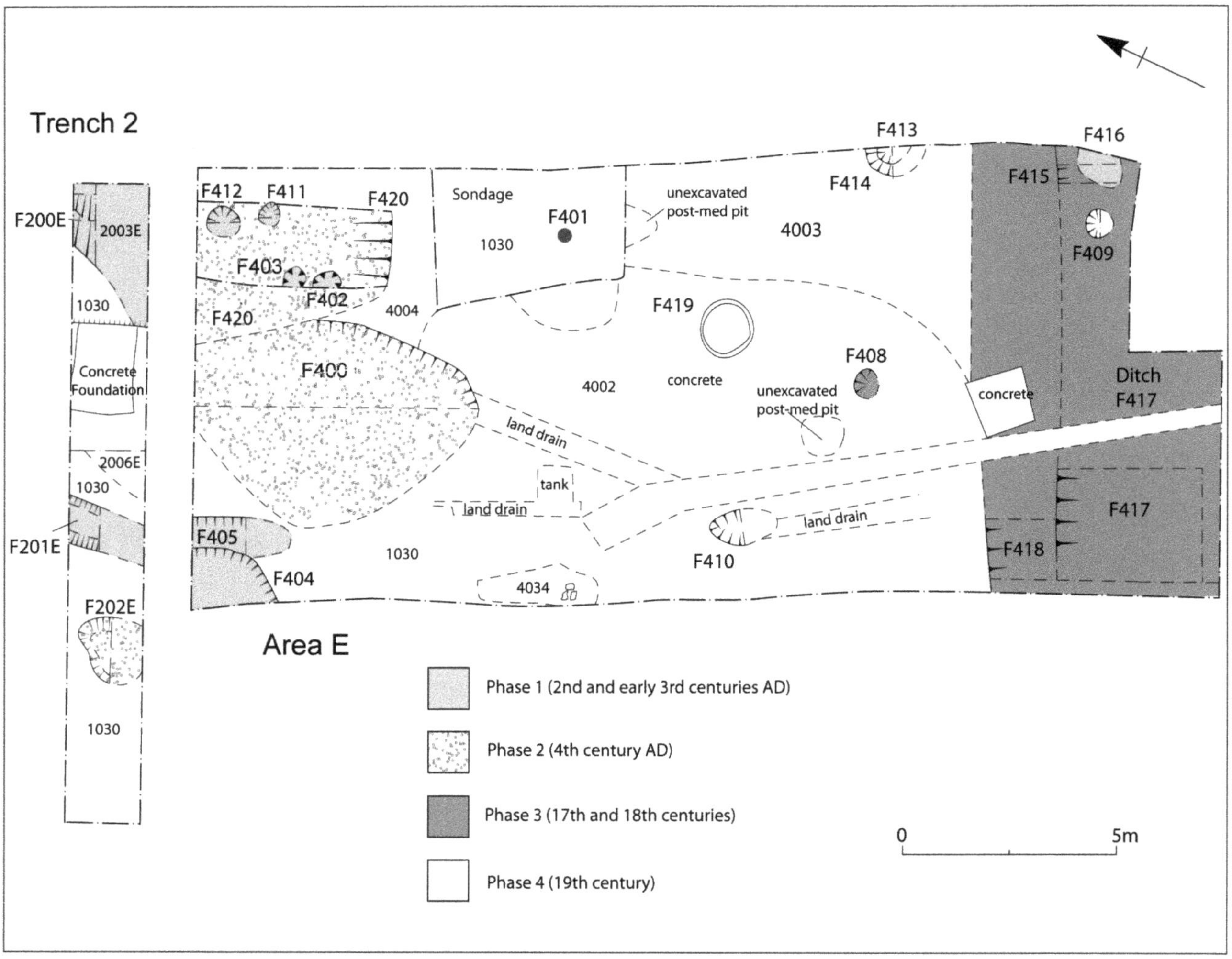

FIGURE 5.7 PLAN OF AREA E AND TRENCH 2

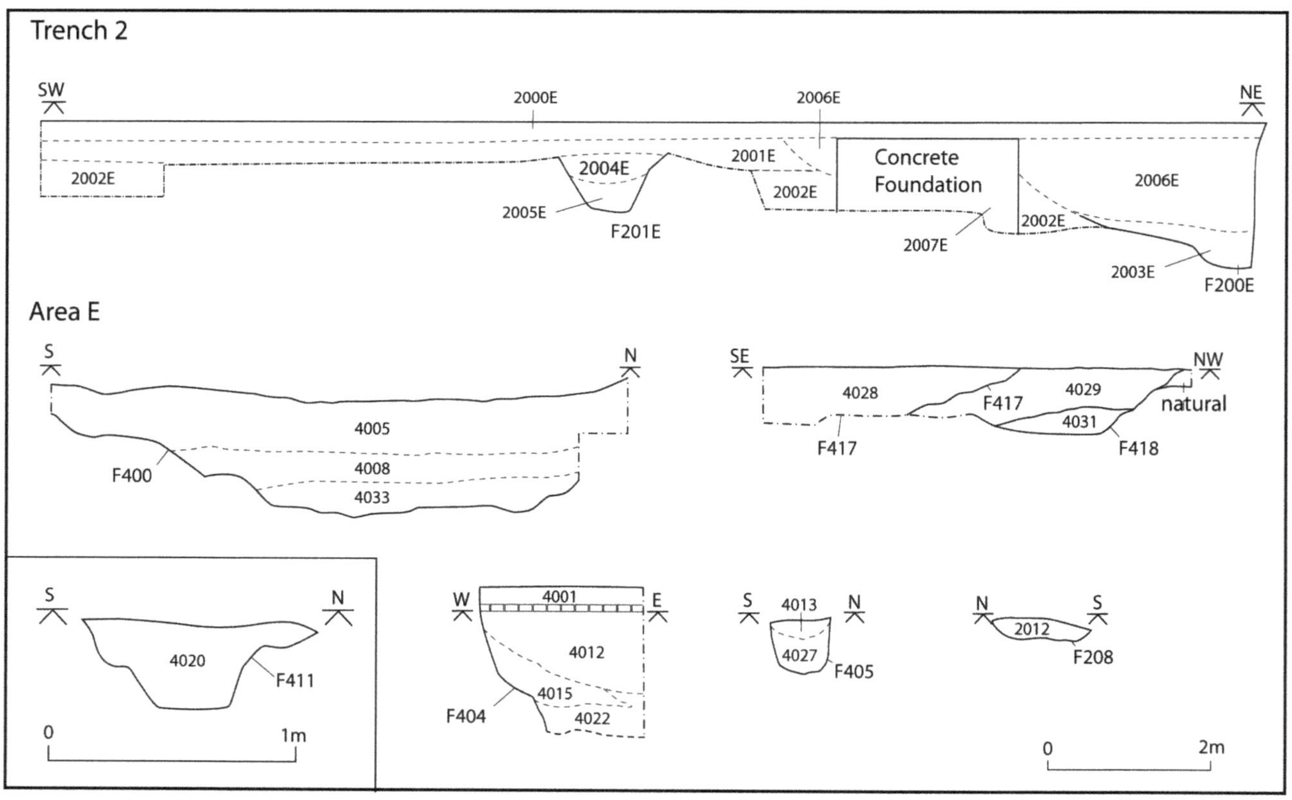

FIGURE 5.8 SECTIONS FROM AREA E AND TRENCH 2

2nd-century, forms. None of the features themselves could be clearly attributed to the 1st century. Activity within Phase 1 was largely focused on pits and shallow ditches, most of which were excavated within Area B. It is possible that these Phase 1 pits result from small-scale quarrying for gravels, and were then filled with rubbish by the early 2nd century. Most of the pits were also fairly shallow, but this may reflect later landscaping and a reduction in the ground level, perhaps during the post-medieval period (see below). The northern extent of the Phase 1 pits would appear to have been marked by a large, east–west aligned ditch in Area A (F107), as no Phase 1 pits were recorded to its north. The ditch may have had a fence or boundary marker (stakeholes F110–F112) to the north. Finds of iron slag in the Phase 1 deposits point to ironworking activity nearby. Of particular note in Area A is a pit (F106) cut into the northern edge of the boundary ditch (F107), which contained an almost complete 2nd-century *tazza*. These have previously been classed as religious vessels, used for burning incense, but have more recently been associated with domestic contexts (see Evans *et al*, this chapter).

Phase 2: 4th century

Area A and Trench 1 (Figs 5.2–5.3)

Much of the northern extent of Area A was sealed by a well-preserved cobbled surface (F100, 1006), generally between 0.13m and 0.20m in depth (Plate 5.3). Parts of this surface comprised smaller stones and slag (1019), which suggests that the surface was subject to occasional repair. A thin layer of light brown clay sand (1022) overlay the surface and produced occasional pieces of animal bone. The cobble surface lay within a distinct hollow that was aligned roughly east–west and had a distinct camber, sloping markedly from south to north, the height varying from 18.76m AOD at the southern most edge of the surface to 18.35m AOD towards the northern edge of excavation, where the cobbles were spread more infrequently. The cobbled surface (1006) was overlain by sand with cobbles (1011) which produced two coins of Constantine I, fragments of mortar and a clay pipe. The northern extent of surface 1006 was overlain by a thin layer of sandy clay (1012) from which late 3rd- to 4th-century pottery forms were recovered including a BB1 jar and Oxfordshire ware mortaria.

To the north of the cobbled surface within Trench 1, were several truncated pits (F100E, F105E, F106E, F107E, F108E), all steep-sided and flat at the base. These survived to between 0.10m and 0.30m in depth and produced sherds dating broadly to the late 3rd or 4th century. One pit (F108E) produced a coin of Constantius II (AD 346–50, Nogueira, this chapter). Pit F108E was cut by F106E which also contained 3rd- to 4th-century pottery.

At the southern edge of Trench 1 was a possible ditch or large pit (F104E), 2.50m in width. This was shallow with a U-shaped profile. The lower fill (1005E) produced a 4th-century penannular brooch, 3rd- to 4th-century pottery sherds and ironworking waste, and the upper fill (1004E) contained tiles in a distinctive Malvernian fabric.

Area E and Trench 2 (Figs 5.7–5.8)

Two large pits (F400 and F420) both survived to a depth of 1.45m. The eastern edge of F400 was cut by F420, which extended beyond the northern and eastern edges

Plate 5.3 Section through the Phase 2 cobbled surface F100 (1006) in Area A

of excavation. The middle and upper fills of F400 (4008 and 4005) and the fill of F420 (4032) produced iron slag, and pottery broadly dating to the late 3rd or 4th centuries. Within Trench 2, pit F202E was cut to a depth of 0.6m, with steep sides and a flat base. While the pit contained pottery broadly dating to the 2nd century, it produced a coin of Tetricus I (AD 270–73, Nogueira, this chapter).

Interpretation of Phase 2

The cobbled surface (F100, 1006) within the hollow way may represent a track in use over an extended period. It seems likely that, during Phase 2, the boundary ditch from Phase 1 (F107) was no longer in use, and pit digging activity resumed across much of the locality, certainly on the eastern half of the site. The sheer size of the pits suggests that they may have originated as gravel quarries, and were subsequently used for waste disposal. The absence of structural features tends to suggest that the locality was predominantly open ground, and this is supported by the environmental evidence (see below).

Phase 3: 17th and 18th centuries (Figs 5.2, 5.4, 5.7 and Plate 5.4)

As shown in Plate 5.4, the Roman features in Area A were truncated by a cut feature (F103) filled with sterile orange clay sand. This was excavated to a depth of 1.20m, but no datable evidence was discovered. It was steep-sided, measured 1.40m in diameter, and it seems most likely that it was a well. The remains of an irregular pit (F109E) were evident below a second pit (F103E) with a third pit (F102E) only visible in section. All contained fragments of post-medieval pottery and cut a brown sand layer (1010E). These pits were sealed by later levelling deposits (1002E, 1018E and 1020E).

The cobble surface (F100, 1006) and layer overlying the surface (1011, 1012 and 1022) were sealed by an organic layer (1005) which filled the east–west aligned hollow. This varied in depth between 0.25m to the south increasing to 0.56m towards the northwestern limit of excavation. It was of particular significance as it provided the largest assemblage of finds, including 1500 sherds of Roman pottery, tile, nails and animal bones, shell, charcoal and slag. Thirteen sherds of samian ware give dates exclusively in the Hadrianic and Antonine periods, whereas three coins of Constantine I give dates between AD 325 and 335. The finds also included bone pins, copper alloy pins, a ceramic counter and a penannular brooch dated to the 2nd or 3rd century (Nogueira, this chapter). However, layer 1005 also produced medieval and post-medieval pottery, tile, clay pipe, glass and brick suggesting a post-medieval context with a significant quantity of residual Roman material.

Within Area B the only evidence for post-medieval activity was recorded in the northern and eastern extent of the area. Pit F210, measuring in excess of 2.40m in diameter and 1.00m in depth, was cut with a steeply sloping profile and may be part of the same feature as F219. In Trench 3 a single pit (F300E) contained brick fragments of possible 18th-century origin.

The natural sand within the central part of Area E was overlain by a levelling layer (4003), which was cut by an east–west aligned ditch (F418). This had a steep profile, and had been recut to the south by a second ditch (F415/ F417). Groundwater ingress at a depth of 18.22m AOD

Plate 5.4 General view of Area A

prevented further excavation. The upper fills of the ditch produced a moderate quantity of medieval pottery sherds and tile (4014). The re-cut (F417) measured in excess of 5.40m wide and, although the profile suggested that F418 and F417 were both part of a substantial defensive system, the southern extent lay beyond the limit of excavation. One further pit (F401) contained 18th-century pottery.

Within Trench 2 was a deep deposit of grey/ black clay silt sand (2006E), which appeared to be the edge of a later levelling deposit or possibly a later ditch. It produced a few artefacts, mostly post-medieval in date.

INTERPRETATION OF PHASE 3

While there is very little evidence for structures in the form of features or architectural finds from Phase 3, there is evidence for several pits and a well. The majority of finds from the fill of the hollow way (layer 1005) date to the 4th century, but the inclusion of medieval and post-medieval material indicates that this is residual and places the context firmly in the post-medieval period. It seems possible, therefore, that many of the finds from 1005 relate to a former Roman midden, levelled during the post-medieval period. This would also be consistent with the low level of abrasion on the Roman sherds (see Evans *et al*, this chapter). Given the lack of evidence for activity between late Roman and late medieval periods, and the implication that the site lay within pasture (see Smith and Tetlow, this chapter) it is not inconceivable that the hollow way remained open, perhaps as a track, after the Roman period,

The east–west aligned ditch (F418) and the recut (F417) in Area E may have resulted from a redefining of Worcester's former extramural Civil War defences. Unfortunately it was not possible to confirm their relationship to the extramural defences through further excavation, due to the proximity of the southern edge of excavation.

PHASE 4: 19TH AND 20TH CENTURIES (Figs 5.2, 5.4, 5.6–5.7)

In Area A, the remains of a north–south aligned wall (F101), constructed of large, worked green sandstone blocks, had been reused as part of the foundations for a later 19th-century wall (F102, 1008), part of the complex of buildings to the northeast of Netherton House. A large sub-rectangular 19th-century domestic and building refuse pit (F105) cut the fill of the hollow way (1005).

In Area B, a series of sub-circular shaped cuts (F206, F209 and F211, F212, F213 and F219) related to a period of extensive building activity. Pits F214, F216, F217 and F218 all cut pit F219. The latest features were brick walls F220 and F221, elements of the foundations of buildings shown on the 1887 Ordnance Survey map extending back from The Butts frontage and arranged around a courtyard.

In Area D, post-medieval and modern building activity consisted of brick footings (F301) and a concrete post (F307), uncovered along the northern edge of the area, and the cut for a fuel tank (F306). Several post-medieval pits were recorded in Area E (F408, F409, F410, F413, F414 and F416) and a well (F419), which was partially excavated to a depth of 18.61m AOD.

In the eastern end of Trench 4 (not illustrated), the natural subsoil (4002E) was encountered beneath 1.75m of overburden. Cutting the subsoil at this depth was a curvilinear feature (F404E), the fill of which was a dark brown silt and rubble mix (4009E) that contained brick, tile, glass, and modern pottery. The remainder of the trench was characterised by walls and cuts likely to be 18th century or later in date. These features were overlain by recent building demolition material measuring between 0.70–0.80m deep and forming the base for the tarmac car park surface.

WATCHING BRIEF (NOT PHASED)

The locations of the plots where a watching brief was undertaken are shown in Fig. 5.1.

PLOT 1 AND AREA F

Across the northern half of Area F and in Plot 1 there was a layer of dark silt (5010) defining a strip, approximately 3m wide, which consisted of a heavy concentration of medium-sized rounded stones. The east–west orientation suggests that this may be a continuation of the Roman track or road/ street identified in Area A. Two distinct areas (5008 and 5009) appeared to cut the southern edge of layer 5010 and may represent patching and repairs.

To the south were several possible postholes/ pits (5002, 5003, 5004, 5005, 5006 and 5007). All of these, except 5003, were filled with brown sandy silt with charcoal flecking and their upper fills contained Roman pottery. The fill of (5003) was characterised by a large amount of broken and burnt stone which may have resulted from deliberate backfilling, but may also suggest industrial activity nearby. The recovery of iron nails and slag would also point to industrial activity in this area. The postholes/ pits and the cobble surface could indicate the presence of a structure on the Roman street frontage.

In the southern half of Area F was a possible pit (5001), fairly large and circular in plan and edged with a band of orange sandy clay (5012). In the southwestern corner of Area F a large cut (F500) was partially exposed in plan. This contained very dark clay sandy silt (5000) which included large lumps of charcoal and a substantial amount of broken post-medieval tile.

PLOT 2

The foundations for Plot 2 were located along the eastern edge of the site and were excavated to a depth of approximately 0.60m. A north–south aligned linear feature (F600, not illustrated) was observed in section only. It contained loose, very dark brown to black sandy silt (6000) which produced two small sherds of Roman pottery along with some 18th-century brown glaze ware. Two large pits (F601 and F602, not illustrated) were also observed in the western facing section of the foundation trench. Pits F601 and F602 were both U-shaped in profile and contained very similar fills. The largest (F601) measured approximately 5m in width by 1.50m in depth and produced Roman pottery mixed with 18th- and 19th-century sherds (6001). The other pit (F602) was only partially exposed in section and measured 3m by 1.5m. The fill (6002) produced a clay pipe stem. A layer of rubble (6003) sealed these features. These features are consistent with the pits and ditches which were uncovered during the excavation of Area E and dated to either Phase 1 or Phase 2.

Plot 3

Several pits (F703, F704, F705 and F706) with U-shaped profiles, and a gully (F702) were observed along the western edge of Plot 3. The gully was aligned northwest to southeast and was filled with dark brown loose sandy-silt (7003) which produced slag and charcoal. Pit F703 was U-shaped and filled with light grey/ brown silty-sand and pebbles (7011). Pit F703 was cut by a second pit (F704) which contained light grey sandy-silt (7012). This was in turn cut by a third pit (F705), which was slightly more irregular with a wide flat base, and was also filled with light grey/ brown sandy-silt (7013). Immediately to the south of these pits was another pit (F706) with a lower fill (7014) of loose, light to mid brown sand-silt, overlain by an upper fill (7015) which was a compact grey-brown silty-sand.

The pits and gully are probably to be associated with the Phase 1 Roman pits and ditches excavated in Area B. All these features were sealed by approximately 1m of compact grey/ brown sandy silt (7010) containing brick, tile and late post-medieval pottery, which may be a post-medieval levelling layer.

The area between excavation Areas B and D contained a deep layer of rubble and demolition debris with large quantities of late post-medieval material, such as clay pipe stems and pottery. This rubble probably resulted from demolition of a building. No archaeological deposits were observed in this area.

Plot 4

The earliest deposit in Plot 4 was very dark brown silt (8002) with a high concentration of pebbles. Two pits (F800 and F801, observed in section only) cut this layer. The earliest of these (F801) was U-shaped in profile and filled by dark brown silt with pebbles (8001). This was cut on its northern edge by a larger pit (F800) which had a rounded profile and was filled with mid brown sandy-silt and redeposited natural (8000). Pit F800 was sealed by a layer of dark brown silt (8003) which, in turn, had been sealed by medium brown sandy-silt (8004). The remainder of Plot 4 consisted of layers of demolition and building debris, which directly overlay the natural subsoil.

Plot 5

On the western edge of Plot 5 an intact sandstone well (F901) of Roman date was uncovered. A sub-oval cut (F900) for the well was observed in plan only and measured approximately 4.6m across. This was filled with brown/ orange silty-sand with pebbles and charcoal flecking (9000), from which a few sherds of Roman pottery were recovered.

The well measured 1.20m in diameter and approximately 6.50m in depth, and was lined with squared red sandstone blocks measuring up to 0.30m x 0.30m. Tool marks were identified on some of the blocks in the upper three courses, but no evidence of mortaring was observed. The well was not excavated below 1m but a core sample from the well showed that, at the base, the first 2m consisted of black, waterlogged deposits which were sealed by 4.50m of loose dark brown silt (9002). This contained Roman pottery, animal bone and slag and probably represents a single episode of backfilling, once the well was abandoned.

No other features were found in the area immediately associated with the well. This perhaps suggests that the well stood apart from other structures on the margins of the Roman settlement, but it is also possible that Roman structures were destroyed by later building activity in the area. The well was sealed by orange/ brown silty-sand (9004) which was overlain by mid brown sandy silt (9005) containing brick fragments. Post-medieval activity in the area is indicated by a post-medieval pit (F903) which cut layer 9005, and which was filled with sandy-silt containing fragments of sandstone (9003).

FINDS

Roman small finds *Isabel Nogueira*

Introduction

The Roman small finds include brooches, pins, bracelets, vessel and furniture fragments, keys, needles, gaming counters, glass and coins (Figs 5.9 and 5.10). They have proved useful in the dating of the site, and have also added to a picture of the daily life of the inhabitants. Many of the Roman small finds can be dated to the 4th century, although some 2nd- and 3rd-century coins were also present, and a penannular brooch is also likely to be of 2nd-century date.

Brooches

1 Ring of a Fowler Type C penannular brooch (Fowler 1960, 152; Crummy 1983, 18). Terminals are coiled and twisted back along the surface of the ring. Circular section. Distorted. Estimated internal diameter 27mm. 2nd to 3rd century. Area A, layer 1005 (hollow way), Phase 3. Fig. 5.9, 1.

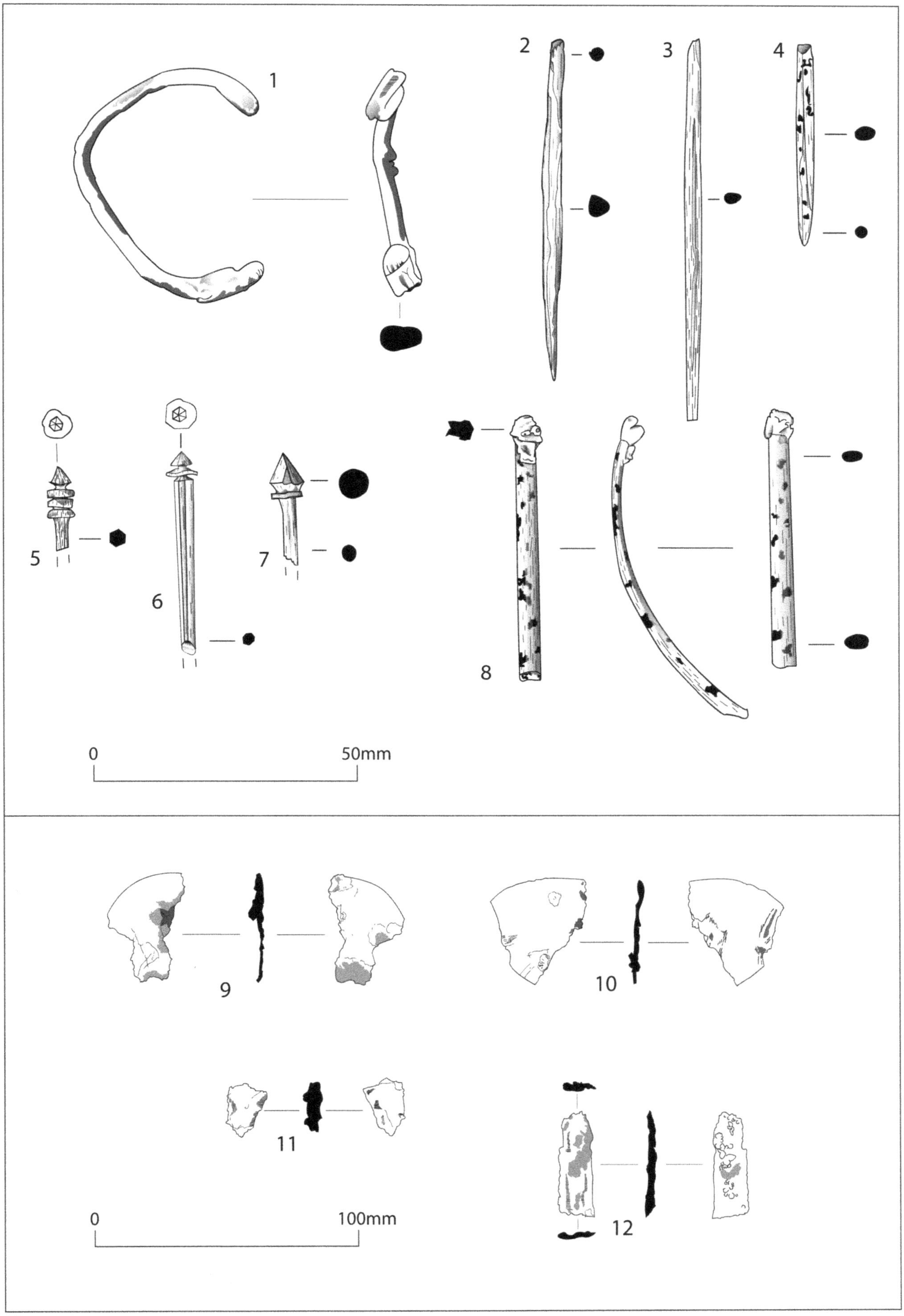

FIGURE 5.9 SMALL FINDS FROM 14–24 THE BUTTS (1–12)

2 Plate of a copper alloy disk brooch. Circular plate, or disc brooch (Alcester: Crummy 1983, 168, fig. 14). 119mm long, 18mm wide, 7mm strip. Area A, layer 1005 (hollow way), Phase 3. Not illustrated.

3 Fragment of iron bow brooch. The fact that only the main body survives (no wings), and the high corrosion to the object make it very difficult to identify with precision. Area A, layer 1005 (hollow way), Phase 3. Not illustrated.

4 Penannular brooch. A flat-hooped penannular brooch, 42mm in diameter and 2mm thick. The terminals are obscured by corrosion products and there is no sign of the pin. The hoop appears to be plain but any decoration may be hidden beneath surface corrosion. The brooch is of a type similar to the extensive group found at Lydney, Gloucestershire (Wheeler and Wheeler 1932). These were all extensively decorated with stamps and engraved lines. Similar examples have been found in Anglo-Saxon cemeteries (eg, White 1988, fig. 5) but it is more likely to be 4th century in date. Evaluation Trench 1, F104E, 1005E, Phase2. Not illustrated.

PINS

1 Bone pin, Type 5. Pin with 1–5 reels beneath a conical or ovoid head (Crummy 1983, 23–24). Faceted section. Swollen shaft. Only head missing. 63mm long. *c* 250 to late 4th or early 5th century. Area A, hollow way layer 1005, Phase 3. Fig. 5.9, 2.

2 Bone pin, Type 5. Pin with 1–5 reels beneath a conical or ovoid head (*ibid*, 23–24). Round section. No head and no tip. Swollen shaft. 71mm long. *c* 250 to late 4th or early 5th century. Area A, hollow way layer 1005, Phase 3. Fig. 5.9, 3.

3 Bone pin, Type 5. Pin with 1–5 reels beneath a conical or ovoid head (*ibid*, 23–24). Faceted section. Swollen shaft broken just below head. 38mm long. *c* 250 to late 4th or early 5th century. Area A, hollow way layer 1005, Phase 3. Fig. 5.9, 4.

4 Bone pin, Type 5a: Pin with 1–5 reels beneath a conical head (*ibid*, 23–24). Faceted section. Three rings. 15mm long – broken below head. *c* 250 to late 4th or early 5th century. Area A, hollow way layer 1005, Phase 3. Fig. 5.9, 5.

5 Bone pin, Type 5a: Pin with 1–5 reels beneath a conical head (*ibid*, 23–24). Faceted section. One ring. 36mm long – broken. *c* 250 to late 4th or early 5th century. Area A, hollow way layer 1005, Phase 3. Fig. 5.9, 6.

6 Bone pin, Type 5b: Pins with 1–5 reels beneath an ovoid head (*ibid*, 23–24). Faceted section. One ring. 22mm long – broken below head. *c* 250 to late 4th or early 5th century. Area A, hollow way layer 1005, Phase 3. Fig. 5.9, 7.

BRACELETS

1 Bone bracelet (MacGregor 1985, fig. 63c). Undecorated. The ends were fastened together by tapering the ends and then riveting them. Greenish-blue spot where the rivet was. 49mm long, 4mm wide and 3mm thick. 4th century. Area A, hollow way layer 1005, Phase 3. Fig. 5.9, 8.

2 Copper alloy bracelet fragment. Group 6 (Crummy 1983, 37–45). Diagonal grooves, meant to imitate cable twisting style (Mould *et al* 2000, fig. 4.8, 47, 111). 3rd century. Area A, hollow way layer 1005, Phase 3. Not illustrated.

3 Plain fragment of a shale bracelet (Woodward and Leach 1993, fig. 129, 13). Round-sectioned. Internal diameter 55mm, 8mm thick. Iron Age to 2nd or 3rd century. Area A, hollow way layer 1005, Phase 3. Not illustrated.

VESSEL FRAGMENTS

1 Base fragments of a copper alloy vessel such as a pan (Manning 1985, 104, plate 50, 32). Possibly a simple lid (cf. Mould *et al* 2000, 120, fig. 4.23, 237). Three fragments, two of which are too small to establish the diameter. 50mm diameter for the largest. 9g; 1g; 1g. 30mm. 4th century. Area A, hollow way layer 1005, Phase 3. Fig. 5.9, 9–11.

2 Base fragments of a copper alloy vessel such as a pan (Manning 1985, 104, plate 50, 32). Possibly a simple lid (cf. Mould *et al* 2000, 120, fig. 4.23, 237). Two fragments, approximately 80mm and 50mm in diameter, 5g and 4g in weight respectively. Both 38mm long. 4th century. Area E, U/ S. Not illustrated.

FURNITURE FRAGMENT

1 Copper alloy decorative strip or appliqué (Cracknell and Mahany 1994, fig. 91, 191). 39mm long. 4th century. Area A, hollow way layer 1005, Phase 3. Fig. 5.9, 12.

KEYS

1 Small copper alloy barb-spring padlock key (Mould *et al* 2000, fig. 4.19, 205; Manning 1985, plate 43, O71). Scrolled loop terminal and strap handle. The bit is missing. Late Iron Age to the 2nd or 3rd century. Area A, hollow way layer 1005, Phase 3. Fig. 5.10, 13.

2 Small copper alloy lift-key (Manning 1985, plate 40, O28). Round-sectioned with a rolled head. Since the bit is missing, it is impossible to determine

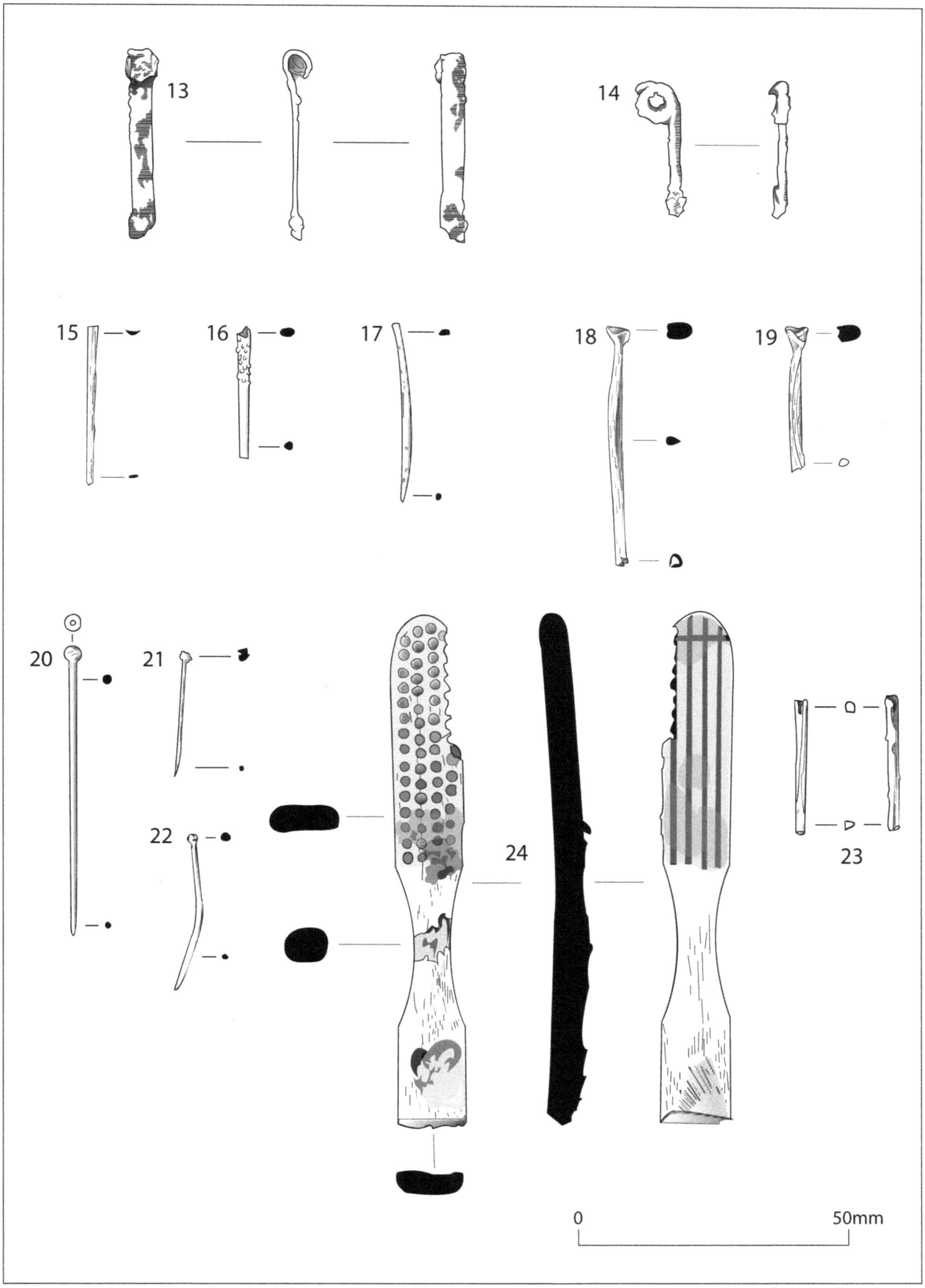

Figure 5.10 Small finds from 14–24 The Butts (13–24)

whether the key was L- or T-shaped. 24mm long. 44; Area A, 1011, F100, Phase 2. Fig. 5.10, 14.

Needles

1 Circular sectioned bone pin or needle point (Bishop 1996, 27, fig. 14, 134). 29mm long; 2mm thick at top, tapering to 1mm at bottom. Common throughout the Roman period. Area A, hollow way layer 1005 , Phase 3. Fig. 5.10, 15.

2 Circular sectioned bone pin or needle point (*ibid*, 27, fig.14, 134). 24mm long; 2mm thick at top, tapering to 1mm at bottom. Has some blue/ green discoloration. Common throughout the Roman period. Area A, hollow way layer 1005, Phase 3. Fig. 5.10, 16.

Gaming counter

1 Baked clay circular counter, probably trimmed from a sherd of oxidized Severn Valley ware (Woodward and Leach 1993, fig. 136, 5–6). 14mm diameter. Used throughout the Roman period. Area A, hollow way layer 1005, Phase 3. Not illustrated.

Glass

1 Eighteen glass fragments were recovered from the Area A, hollow way layer 1005, Phase 3.

The most interesting piece was a convex fragment of colourless glass cup with straight or curved rim (Price and Cottam 1998, 117–19, fig. 48a). Decorated with a 'festooned' trail. Usually rim diameters for this type of vessel are approximately 70–90mm, and the height can vary from 65–80mm. This fragment is 27mm long. Very common vessel during the late 3rd to late 4th century. Area A, hollow way layer 1005, Phase 3. Not illustrated

Coins

The coins are catalogued in chronological order as follows: Issuer; denomination; date; reference; mint; description of obverse; description of reverse; weight (g); diameter (mm); wear; find number; feature number; context number; area. Condition of the obverse and reverse is denoted by the following abbreviations: uw – unworn, sw – slightly worn, w – worn, vw – very worn, ew – extremely worn and c – corroded.

1 TRAJAN; Dupondius; AD 101–2; Robertson 1971, plate 7, 231; Rome
Obverse: Head of Trajan, bust radiate right. Slight drapery on left shoulder. IMP NERVA CAE [TRAIA]N [AVG] GERM PM
Reverse: Justitia, draped, seated left on chair with cornucopia in arms, holding sceptre in right hand, left hand on hip, fall of drapery falling over lap. TRPO[T C]OS [.]II[II.P.]
(8g; 26mm; w/ w; 38; F207; 2011; B; Phase 1).

2 SEPTIMIUS SEVERUS OR FAMILY; Denarius; AD 193–211
It was not possible to further identify this coin. It may have been overstruck, or a copy
(3g; 17mm; vw/ vw; 39; F200; 2004; B; Phase 1).

3 SEPTIMIUS SEVERUS; Sestertius; AD 195; Hill 1964, plate 18, 163; unknown mint
Obverse: Head of Septimius Severus, laureate, right. [L. SEPT. SEV. PERT. AVG. IMP. V]
Reverse: Palladium/ Minerva standing left holding spear and shield. [P. M. TR. P. III COS. II P. P]

(2g; 17mm; ew/ ew; 41; F417; 4014; E; Phase 3).

4 Tetricus I; Antoninianus; (270–73) Mint unknown
Obverse: [IMP CAES TETRIC-]VS[PFAVG] Bust r, wearing radiate crown
Reverse: Obscured by corrosion
Corroded and partly obscured by a concreted stone, (3g; 21mm; w/ c; Die axis unknown; F202E; Trench 2; Phase 2).

5 Constantius II; Follis (346–50) As Carson and Kent 1960, 69, plate II, 1133.
Obverse: DN CONSTANTIVS PF AVG Mint mark not on flan
Reverse: FELTEMP-REPARATIO Phoenix
(2g; 17mm; w/ w; Die axis 6; F108E; 1017E; Trench 1; Phase 2).

6 CONSTANTINE I; Follis; AD 313–17; Robertson 1982, plate 46, 97; Arles
Obverse: Bust r. of Constantine I, laureate, draped, cuirassed. IMP CONSTANTINVS P [F AVG]
Reverse: Sol, radiate, naked except for cloak over left shoulder, standing right, head left, raising right hand and holding globe in left hand. [SOL] IN[VIC]TO COMITI
(4g; 21mm; vw/ w; 35; F100; 1011; A; Phase 2).

7 CONSTANTINE I; Follis; AD 325–30; Robertson 1982; unknown mint
Obverse: Head of Constantine I, laureate, right. CONS[TANTINVS AVG]
Reverse: A fortified camp-gate, two turrets, star above. [PROVIDENTIAE AVGG]
(3g; 19mm; vw/ ew; 13; 1011; A; Phase 2).

8 CONSTANTINE I; Follis; AD 330–35; Robertson 1982, plate 46, 92; Lyon, PLG
Obverse: Bust of Constantine I rosette-diademed, draped, cuirassed, right. CONSTANTI NVS MAX AVG
Reverse: Two soldiers, cloaked, standing facing one another, each holding spear, other hand on shield. Between them are two standards. GLOR IA EXERC ITVS
(2g; 14mm; w/ vw; 27; hollow way layer 1005; A: Phase 3).

9 CONSTANTINE I; Follis; AD 330–35; Sutherland 1974, plate 267, 533; Unknown mint
Obverse: bust of Roma, helmeted and draped, facing left; VRBS ROMA
Reverse: too corroded to identify
(1g; 18mm; ew/ c; 28; hollow way layer 1005; A; Phase 3).

10 CONSTANS I; Follis; AD 337–40; Robertson 1982, plate 65, 6; Trier P
Obverse: Bust of Constans, rosette-diademed, draped, cuirassed, right. [CONSTAN S P F AVG]
Reverse: Two victories, draped, standing facing one another, each holding wreath and palm. [VICTORIAE DD AVGGQ NN]
(1g; 13mm; vw/ vw; 53; hollow way layer 1005; A; Phase 3).

DISCUSSION OF ROMAN SMALL FINDS

With the notable exception of a number of the coins, the majority of the Roman finds came from the Phase 3 hollow way fill (1005) which overlies the cobbled Roman street in Area A. This layer also contained medieval and post-medieval artefacts and probably comprises material which was redeposited as part of post-medieval levelling activity.

Ten identifiable coins are Roman in date, ranging from the earliest at AD 101–2 to the latest at AD 337–40. The majority (60%) of the Roman coins date from the 4th century. The earliest is a brass *dupondius* issued by Trajan between AD 101 and 102, found in a small gully (F207) in Area B, and it helps to date that feature to Phase 1. Also attributable to Phase 1 in Area B is a silver *denarius* issued by Septimius Severus or family (AD 193–211) found in the fill of rubbish pit F200. A bronze *follis* issued by Constantius II (AD 346–50) was found in a pit (F108E) in Trench 1, and helps to date this feature to Phase 2.

The remaining Roman coins were found in post-Roman Phase 3 contexts and are regarded as residual or redeposited finds. They include another coin issued by Septimius Severus, this one from AD 195 with Minerva/ Palladium on the reverse (Hill 1964, 18), and five coins from Area A all issued by the House of Constantine between AD 313–40; four by Constantine I, and one by his son Constans I. The House of Constantine coins are all generally common bronze *follis* types, and are quite typical of 4th century Roman sites.

Penannular brooches are common to Roman sites in Britain, and the ring of the penannular brooch found in the hollow way fill (1005) belongs to a Fowler type C (Fowler 1960, 152; Crummy 1983, 18). It seems to be early in date due to its simplicity. The design of such simple penannular brooches can be tentatively dated to the 2nd century (Barker *et al* 1997, 155), although it may have been based on an earlier, pre-conquest model (Crummy 1983, 19). Brooches were often kept for long periods of time (Cracknell and Mahany 1994) and could be adapted and enhanced with decoration (Crummy 1983, 19). A further fragment from a plate brooch could not be accurately identified; however, it seems clear that plate brooches date mainly to the 2nd century, although the type may have continued into the 3rd century (Barker *et al* 1997, 155; Cracknell and Mahany 1994, 168).

Brooches seem to have lost their popularity to bracelets as the main form of personal adornment during the late Roman period (Woodward and Leach 1993). Only one fragment of a copper alloy bracelet was discovered, which was found to belong to Crummy's group 6 (Crummy 1983, 37–45), dating to the 3rd century. Characteristic decoration for this group comprises mainly diagonal grooves on the outside of the bracelet to imitate the previously popular style of 'cable-twisting' (Mould *et al* 2000, fig. 4.8).

The only fragment of a possible shale bracelet retrieved from the site is plain and undecorated, which correlates with other Roman sites in Britain. Previous work on contemporary sites has shown that more than half the bracelets present were plain, as they were during the Iron Age (Lawson 1976, 166). These were in use from the 1st century BC to the 2nd or 3rd century. The worked bone bracelet dates to the 4th century, when these were quite common. It is plain and undecorated, and the ends held together by rivets (MacGregor 1985, 112).

All six bone hairpins retrieved have been attributed to Crummy's (1983) Type 5, which range in date from the mid 3rd century to the late 4th or early 5th century. Crummy sub-divides this type according to the shape of the pin head; subtype A is conical with grooves, while in subtype B the head is more of an ovoid, or flame-shape. On this basis two of the pins can be attributed to subtype A, and a third to subtype B, whilst fragments which comprise only of the shaft cannot be further categorised.

The two needle points resemble Crummy's types 1 or 2, both of which were common throughout the Roman period (1983, 65–67; Barker *et al* 1997, 117). One of the needle points has traces of a blue/ green substance, suggesting that it may either have been in contact with metal or that dye may have been used to decorate the surface.

ROMAN IRON OBJECTS *Erica Macey-Bracken*

A quantity of other Roman iron objects was also recovered from the site. The assemblage was generally very corroded, and many items were difficult to identify.

Two iron objects and six nails were recovered during the evaluation. The objects were a cylindrical iron bar (Trench 1, F104E, 1005E, Phase 2) and a thin crumpled amorphous iron sheet (Trench 1, F100E, 1002E, Phase 3). Three of the nails were of indeterminate form, but the others were smaller than 150mm in length and of Manning Type 1b (Manning 1985, 133–35). They were from Phase 2 and Phase 3 contexts (see Table 5.1).

Context	Phase	Type 1a	Type 1b	Type 3	Type 4	Type 7	Indeterminate
2000 EVALUATION							
1002E	3		2				
1005E	2		1				
2004E							1
2008E							1
3007E							1
2003 EXCAVATION							
1004	4	1					1
1005	3	5	12	1	1		6
1011	2					1	
1012	2	1					
1017	1	1					
2004	1	1					1
2010	4						1
2013							3
2014	3			1			
2022	4						1
3004	1						1
4003	3			1			
4005	2	1					
4016	3		1				1
4025	3		1				

TABLE 5.1 LOCATIONS AND TYPES OF NAILS

The bulk of the assemblage recovered from the excavation comprised iron nails, with a total of 47 nails being recovered. Most of the nails were near complete, and ten were of Manning's (1985) Type 1a, larger than 150mm in length, and 15 were of the smaller Type 1b. Type 1 nails are the most common Roman nail form, found on the majority of Roman sites. Three examples of Manning's Type 3 were also recovered, as was a single example of Type 4. It is possible that some of the indeterminate nails may be of more recent date, particularly those recovered from contexts dating to later phases of occupation on the site. A small rivet or stud is probably a Manning (1985) Type 7 nail, commonly called a tack. The frequency, location and phase of the various nail types can be seen in Table 5.1.

Other iron items from the site were much harder to identify, due to the presence of corrosion products, which in many cases obscured the original shape of the items. Among the more identifiable items were nine fairly flat pieces of iron (Area A, hollow way layer 1005, Phase 3, x 4; Area A, layer 1012, Phase 2 x 2, Area B, F210, 2014, Phase 3, x 3), which may have come from strips of iron, such as those that were used to bind wooden boxes. One of the pieces (1005) has a right-angled bend, and may have belonged to the corner of such a strip.

POST-ROMAN SMALL FINDS *Isabel Nogueira*

NEEDLE

1 Copper alloy needle (Bishop 1996, fig. 15, 148, 28). Round sectioned. 32mm long, 2mm thick. Medieval or post-medieval. Area E, F417, Phase 3. Fig. 5.10, 17.

'PENS'

1 'Pen' made from bird fibula (MacGregor *et al* 1999, 1976, fig. 932, 7976 and 8059). Possibly a 'parchment pricker'. 44mm long, broken. Medieval. Area A, layer 1005 (hollow way), Phase 3. Fig. 5.10, 18.

2 'Pen' made from bird fibulae (MacGregor *et al* 1999, 1976, fig. 932, 7976 and 8059). Possibly a 'parchment pricker'. 26mm long, worked end missing. Medieval. Area A, hollow way layer 1005, Phase 3. Fig. 5.10, 19.

DRESS PINS

1 Copper alloy dress pin (Barker *et al* 1997, 194, fig. 296, 1–2). Wound-wire head. 53 mm long, head is 3 mm, shaft 1.5 mm. 4th century onwards. Area A, layer 1002, Phase 4. Fig. 5.10, 20.

2 Copper alloy pin (Neal *et al* 1990, 137, fig. 130, 347). 22 mm long, 2 mm wide for head, 1 mm wide for shaft. Medieval or post-medieval. Area E, F416, 4026, Phase 3. Fig. 5.10, 21.

3 Copper alloy pin (Neal *et al* 1990, 137, fig. 130, 347). 26 mm long, 2 mm wide for head, 1 mm wide for shaft. Medieval or post-medieval. Area E, F416, 4026, Phase 3. Fig. 5.10, 22.

4 A copper alloy (brass?) dress-maker's pin 33mm long with a simple domed head. Modern-looking artefact that was perhaps intrusive. Evaluation Trench 1, F104E, 1005E, Phase 2. Not illustrated.

EARRING

1 A hoop of thin copper alloy wire 25mm in diameter and 1mm thick. Possibly a simple earring. Evaluation Trench 2, F200E, 2003E, Phase 1. Not illustrated.

CHAPE

1 Copper alloy lace chape, (Egan and Pritchard 1991, 281–90). 24mm long. Medieval or post-medieval. Area E, F417, 4014, Phase 3. Fig. 5.10, 23.

BRUSH

1 Bone compound brush (MacGregor 1985, fig. 99). Blue/ green stains both in the front and the back where the tufts would have been. 'Has the characteristic funnel-shaped perforations linked on the back by grooves to accommodate the wires which anchored the tufts' (*ibid*, 183). 930mm long. Most are dated to the 17th century or later. Area B, F209, 2013, Phase 4. Fig. 5.10, 24.

DISCUSSION OF POST-ROMAN SMALL FINDS

Two 'pen' fragments from hollow way layer 1005 are unlikely to date to the Roman period. These are usually made of goose radii, but can also be made from most bird fibulae, and their use is still not certain. Some have suggested that they may have been used as 'parchment prickers', or as pens used only for drawing simple lines (MacGregor 1985). One of their ends would be obliquely cut and sharpened, but neither of the two examples retrieved have that end present. These probably date to the medieval period and were most likely used by scribes to simplify the task of writing in straight lines (*ibid*, 125).

F209 produced a bone brush, which dates to the 17th century or later (*ibid*, 183). Compound brushes of this type are usually made from the long bones of cattle, although more recent examples were often made of baleen. The tufts could have been made of bristle, but some examples, such as the one from F209, show green stains 'from copper salts, indicating that the tufts were of fine wire' (*ibid*, 183). Some of the possible functions for these objects include clothes brushes and toothbrushes, depending on the size.

JETTON *Roger White*

Copper Alloy jetton (Trench E SF1). Flange incomplete but 80% of the disc survives.

Obv: Four *fleur de lys* arranged within lozenge and quatrefoil made of overlapping arcs with three pellets above each intersection of the arcs. Solid border with, around edge from initial quatrefoil, [--]TT GOTT [GO]TT T

Rev: Saltire in lozenge with *fleur de lys* in angles within cusped quatrefoil. Solid border with indistinct legend perhaps similar to that seen on obv.

The obverse is much crisper than the reverse suggesting the use of a very worn die for the reverse. The design is not dissimilar to French jettons of the second half of the 16th century (http://www.ukdfd.co.uk/ukdfddata/showrecords.php?product=28786&cat=84&limit=recent&date=1121510760). This is a typical find for a later medieval site.

CLAY PIPE *Erica Macey-Bracken*

A total of 41 pieces of clay pipe were recovered from the site: 1002, Phase 4 x 5; 1005 (hollow way), Phase 3 x 1; F100 (1011), Phase 2 x 1; F105, (1013), Phase 3 x 1; F210, (2014), Phase 3 x 6; F210, (2022), Phase 4 x 12; F210, (2026), Phase 3 x 2; 4003, Phase 3 x 5; F407, F415 (4025), Phase 3 x 2; F417, (4014), Phase 3 x 4. Most of the pieces recovered were stem fragments, including a mouthpiece (1002), although two bowl fragments were also recovered.

The mouthpiece was notable in that it was made of a very pale terracotta-coloured clay and was probably of German manufacture. Short-stemmed pipes of this type were especially popular during the early 20th century and they continued to be imported until WWII. The curved form is also typical of this type of pipe, which was used by adults for smoking and by children for blowing bubbles (Dr David Higgins, pers. comm.).

The most complete bowl fragment (2014) is of probable mid to late 17th-century date (Ayto 1999, 8, bowl no. 5) and has a very faint roulette pattern at the top of the bowl. The other bowl fragment is too small to be of any diagnostic use.

MISCELLANEOUS FINDS *Erica Macey-Bracken*

As with most of the rest of the finds, these items were fragmentary, although individual pieces were largely unabraded. The finds were quantified by count and weight and scanned macroscopically for the purposes of this report. A small quantity of charcoal was recovered from Phase 1 (Area B, F200, 2004) and Phase 3 (Area A, 1005) and a flint flake, with evidence of retouch on both sides, came from context 3007. A glass bead from Phase 3 (Area A, hollow way layer 1005) was of opaque green glass and measured only 3mm in diameter and another from Phase 4 (Area A, layer 1002) was much larger, at 8mm in diameter, and was made from opaque blue glass. A small piece from a wooden rod or dowel was unstratified.

METALWORKING REMAINS *Eleanor Blakelock*

INTRODUCTION

Excavations at several Roman sites in Worcester have discovered a significant industrial component. At Deansway 500m to the south, ironworking waste products were dated to the 2nd to 3rd century (Dalwood and Edwards 2004, 43–44). In the 3rd and 4th centuries, the focus of industrial activity south of The Butts shifted from Deansway to

Context	Feature	Area	Total Waste (g)	Type	Phase	Date
2004	200	B	1398	Ditch Fill	1	2nd–3rd century
3004	304	D	1937	Ditch Fill	1	2nd–3rd century
3005	304	D	1232	Ditch Fill	1	2nd–3rd century
4007	402	E	6234	Pit Fill	1	2nd–3rd century
4012	404	E	11427	Pit Fill	1	2nd–3rd century
4015	404	E	10620	Pit Fill	1	2nd–3rd century
1011	100	A	2139	Cobble matrix	2	4th century
1012	-	A	1317	Layer	2	4th century/ post-Roman
4005	400	E	13021	Pit Fill	2	4th century
4008	400	E	3992	Pit Fill	2	4th century
1005	-	A	5262	Layer	3	17th–18th century
4003	-	E	3088	Layer	3	17th–18th century
4016	417	E	2659	Ditch fill	3	17th–18th century

TABLE 5.2 CONTEXTS CONTAINING MORE THAN 1KG OF IRONWORKING WASTE

Context	Feature	Date/ Century	Tap Slag	Run Slag	Smelting Slag	Vitrified Lining	Ore	Fuel	Undiag. Slag	Iron
1003		4th							21	
1004		19th								83
1005		4th +	1285	76	2063	278	403		488	669
1011		4th +	140		1662				69	268
1012		4th +	536		718				63	
1018	108	2nd–3rd	18							
2004	200	2nd–3rd	873		372				153	
2005	201	Unknown	13							
2006	202	2nd–3rd	4							
2007	203	2nd–3rd	60							
2008	204	2nd–3rd	57							
2009	205	Roman	119							
2011	207	Early 2nd	393							
2012	208	Roman	10							
2014	210	Post-Roman	21							
3009	303	Late 2nd	21		118					
3004	304	Late 2nd	1918						19	
3005	304	Late 2nd	1105			78			49	
3006	305	Late 2nd	118		795					
4003		Post-medieval	615		2044			10		419
4005	400	3rd–4th	5574	121	6371	324		8	623	
4007	402	Roman	2972		2854				408	
4008	400	4th	2116	49	1827					
4012	404	Late 2nd	6481		4177	119			598	52
4013	405	2nd	30		248					
4014	406	Post-medieval	238		558					
4015	404	Late 2nd	4317		4255				2048	
4016	407	Post-Roman	1552	257	837				13	
4020	411	2nd–3rd	8							
4021	412	2nd–3rd	83							
4022	404	Late 2nd	124							
4025	415	Post-medieval							17	
4026	416	Post-medieval	29							
4032	420	4th	32							
U/ S		Unknown	69						330	

TABLE 5.3 QUANTITIES (IN GRAMS) OF DIFFERENT TYPES OF IRON WASTE RECOVERED BY CONTEXT

Broad Street (see Fig. 5.1), where several furnaces were recorded. Further 3rd- to 4th-century ironworking waste was also recorded adjacent to The Butts at Farrier Street (*ibid*, 101).

The evidence for metalworking at 14–24 The Butts comprises ironworking waste recovered mostly from deposits in pits and ditches. The stratigraphy and dating of the site (see above) indicates that, in most of these deposits, the ironworking waste is likely to be Roman in date. Some contexts are dated to the post-medieval period (Phase 3), but also contain residual Roman pottery and other Roman finds. This suggests that the ironworking waste in these deposits could also be residual Roman material. Thus, whilst there is certainly evidence for Roman period ironworking from the site, there is no unequivocal evidence for any post-Roman ironworking. No trace of any ironworking structures was found, but this does not preclude the possibility of their former existence. Much of the Roman land surface was heavily truncated in post-Roman times, and this could have removed traces of Roman structures.

IRONWORKING PROCESSES AND CATEGORIES OF IRONWORKING WASTE

There are broadly two processes involved in ironworking: smelting (extracting metal from the ore), and smithing or forging (shaping the object). Both create different kinds of waste that can often be distinguished on the basis of their morphology, as described below.

Iron smelting took place in bloomery furnaces, which were typically clay-built, rounded structures in the Roman period. Iron ore was fed into the furnace where it reacted to create a spongy mass of iron metal known as a bloom. The waste from this process formed a liquid slag that was often tapped from the bottom of the furnace (Bayley *et al* 2001).

The ironworking waste from The Butts was classified predominantly using Bayley *et al* (2001). The categories comprise tap slag, runs, smelting slag, vitrified lining, ore, fuel, iron objects and undiagnostic slag. There is a summary of the main contexts producing ironworking debris in Table 5.2, with a description of the debris by context in Table 5.3.

Tap slag and runs were by-products of the smelting process, produced by running slag out of the furnace (tapping) when it was hot and fluid. This waste had a characteristic shape, resembling a flow of lava, and the lower surface may be rougher than the upper one as it comes into contact with the ground. This slag type was not porous and was dark grey in colour. Large numbers of the tap slag and run fragments appeared to be tubular in form.

Smelting slag consisted of large blocks of slag, often with fuel impressions in the surface. These were a similar colour to the tap slag. It had obviously been fluid within the furnace, but did not look as though it had flowed in the same way as the tap slag. The porosity of this slag varied greatly.

Vitrified lining consisted of small fragments of quartz-rich clay that had been subjected to heat. The outer surface was orange/ red with a black inner (vitrified) surface. This is most likely to be part of a furnace or hearth structure and the vitrified surface is most likely from the inside of the furnace. Some fragments of vitrified lining had slag adhering to them.

Undiagnostic slag did not have sufficient characteristics to be categorised; similar materials may be produced by either smelting or smithing operations.

DESCRIPTION AND CLASSIFICATION OF THE IRONWORKING WASTE

A total of 67.9 kg of ironworking debris was recovered from the site. Figure 5.11 shows the relative proportions of the different categories of waste. Over 89% of the waste from the assemblage was tap slag, runs and smelting slag, all produced during iron smelting, and therefore it can be concluded that smelting was the main ironworking process taking place. There was no evidence of smithing activity at the site as no slag with the characteristic morphology typical of smithing slag was identified. Vitrified lining can be produced by both the smelting and smithing processes, but since no smithing slag was found, it is most likely to be furnace lining.

In Table 5.2, contexts containing more than 1kg of metalworking debris are listed. The waste from these contexts comprises over 94% by weight (wt %) of the debris recovered from the site. Of this waste, over 75wt % was from Roman dated contexts, while only 10wt % came from post-Roman contexts and the remainder was undated. Therefore, the ironworking activity probably took place largely in the Roman period. As stated above, whilst some of the contexts appear to be post-Roman, the

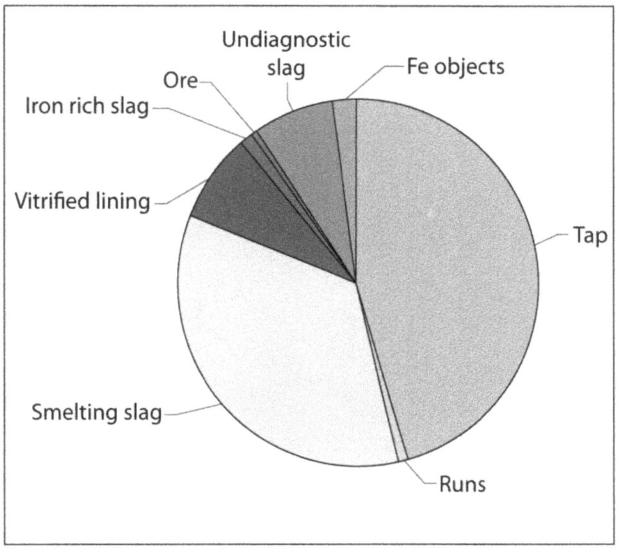

FIGURE 5.11 PROPORTIONS BY WEIGHT OF DIFFERENT TYPES OF IRONWORKING WASTE

PLATE 5.5 IRON ORE FRAGMENT FROM 1005

evidence suggests that it is more likely to be redeposited Roman waste than evidence for post-Roman ironworking. Most of the metalworking debris came from Trench E, to the east of the site, in pits that were dated to the 3rd to 4th century. The contents of these pits are similar to the debris found during the excavation of Farrier Street (Dalwood *et al* 1994).

Material other than slag or vitrified lining found in the assemblage included one piece of iron ore (Plate 5.5), two pieces of charcoal and a rounded quartz pebble that could have been used as a hammer stone. A few iron objects were also found in the assemblage including a piece of unrefined iron, probably a fragment of bloom.

RESEARCH QUESTIONS

Previous excavations at Deansway, Worcester have revealed similar debris, and analytical studies of that ironworking debris have been carried out. Samples from The Butts were analysed to provide a comparative assemblage for the Deansway material. The source of the iron ore for the Roman ironworking industries at Worcester was also investigated.

ANALYTICAL METHODS

Pieces of tap slag, smelting slag, ore, furnace lining and probable bloom were chosen from contexts that produced large amounts of iron-smelting waste. Small samples were taken from each piece and mounted in epoxy resin. They were ground and polished to a 1μm finish and then coated with carbon. Examination and analysis was carried out using a scanning electron microscope (SEM) in both backscatter and secondary electron mode. Both modes were used to assess the condition and homogeneity of the samples. The composition of the samples was determined using an energy dispersive spectrometer (with Germanium detector) attached to the scanning electron microscope and calibrated with a cobalt standard. Spectra were collected at 25kV and 2nA for 150 seconds live time. The spectra were quantified using the Oxford Instruments SEMQuant software. A small area (typically 3mm by 2mm) was analysed on each sample. Slag is heterogeneous, often with a coarse microstructure, and small areas may not be representative of the whole sample. To reduce this

No	Context	Waste Type	Na_2O	MgO	Al_2O_3	SiO_2	P_2O_5	K_2O	CaO	TiO_2	MnO	FeO
1	4012	Vitrified	1.1	1.5	13.6	74.4	0.2	2.6	0.1	0.7	<0.1	5.8
		lining	3.6	0.6	13.0	76.0	0.2	2.3	0.4	0.5	<0.1	3.4
			0.8	0.9	9.8	80.0	0.2	2.3	0.2	0.7	<0.1	5.0
2	1005	Ore	<0.5	0.4	0.4	1.6	<0.1	<0.1	<0.1	<0.1	<0.1	97.2
			<0.5	0.5	0.3	1.3	<0.1	<0.1	0.1	<0.1	<0.1	97.5
			<0.5	0.5	0.5	1.7	0.1	<0.1	0.1	<0.1	<0.1	96.9
3	4012	Tap	<0.5	0.9	6.9	30.6	0.2	1.8	2.1	0.2	0.2	56.7
			<0.5	0.8	6.9	30.8	0.2	1.7	2.1	0.2	0.1	56.6
			<0.5	0.9	7.4	30.9	0.3	1.4	1.8	0.3	0.2	56.4
4	4005	Smelting	<0.5	1.0	4.3	18.2	0.2	1.1	1.3	0.2	0.2	73.2
		slag	<0.5	1.1	4.6	18.3	0.2	1.1	1.3	0.1	0.2	72.7
			<0.5	1.0	4.7	18.8	0.3	1.0	1.1	0.1	0.2	72.6
5	4005	Tap	<0.5	1.0	5.1	24.9	0.3	1.0	1.2	0.2	0.1	65.9
			<0.5	0.9	5.3	25.4	0.3	1.0	1.2	0.2	0.1	65.2
			0.5	0.7	5.7	25.3	0.3	1.1	1.4	0.2	0.1	64.5
6	1005		<0.5	1.2	4.6	20.1	0.7	0.7	1.2	0.2	<0.1	71.0
		Slag around	<0.5	1.2	4.6	20.2	0.1	0.7	1.2	0.2	<0.1	71.4
		bloom	<0.5	1.6	4.0	19.0	0.3	0.6	1.4	0.2	0.1	72.3

TABLE 5.4 CHEMICAL COMPOSITION OF THE IRONWORKING WASTE SAMPLES (NORMALISED)

	Na2O	MgO	Al2O 3	SiO2	P2O5	SO3	K2O	CaO	TiO2	MnO	Fe2O3
Standard A	14.93	2.78	1.04	69.75	0.14	0.10	3.00	5.25	0.82	1.04	1.14
SEM Analysed A	15.2	2.7	1.0	69.1	0.1	0.2	3.2	5.2	1.0	1.1	1.1
Standard B	17.96	1.09	4.61	64.98	0.87	0.53	1.10	9.04	0.09	0.26	0.36
SEM Analysed B	17.5	1.1	4.5	63.7	0.9	0.7	1.2	9.6	0.1	0.3	0.4
Standard D	1.24	4.07	5.47	56.42	4.06	0.31	11.67	15.28	0.39	0.57	0.54
SEM Analysed D	1.5	4.1	5.4	55.9	4.1	0.3	11.7	15.5	0.5	0.6	0.5

TABLE 5.5 CHEMICAL COMPOSITION OF THE CORNING GLASS STANDARDS COMPARED TO THE ANALYSED RESULTS (NORMALISED, AVERAGE OF THREE)

	Na2O	MgO	Al2O3	SiO2	P2O5	SO3	K2O	CaO	TiO2	MnO	FeO
MiniProconv A122	0.67	0.40	7.38	24.03	0.26	0.07	1.10	1.00	0.33	3.09	61.68
Sandvik conv A123	0.70	0.40	7.33	25.51	0.26	nm	1.00	1.41	0.31	3.11	59.96
SGAB2 ICP A226	0.45	0.47	7.80	25.50	0.31	nm	1.07	1.75	0.33	3.21	59.13
SGAB2 ICP A226	0.56	0.43	7.49	25.50	0.25	nm	1.10	1.58	0.34	3.32	59.42
SGAB2 ICP A226	0.56	0.39	7.67	25.87	0.52	nm	1.07	1.57	0.43	3.14	58.79
Average Slag Standard	0.59	0.42	7.53	25.28	0.32	-	1.07	1.46	0.35	3.18	59.79
EDS analyses	1.4	0.3	8.0	21.2	0.3	0.2	1.0	1.4	0.3	3.0	62.9
	1.5	0.5	8.5	21.6	0.2	0.3	1.0	1.5	0.3	3.0	61.5
	1.3	0.4	8.0	21.8	0.2	0.3	1.0	1.4	0.3	3.1	62.2
Average EDS Analyses	1.4	0.4	8.2	21.5	0.3	0.3	1.0	1.5	0.3	3.0	62.2

nm = not measured

The slag standard is from the Viking Age iron production site at Gryssen in Dalecarlia (Kresten and Hjärthner-Holdar 2001). The published original analyses of the standard using different techniques by nine laboratories resulted in different compositions, some of which had low overall totals. The six results selected from those published, to determine an average with which to compare, were those with totals closest to 100% and which included all the oxides of interest. Results that varied considerably from those of the other laboratories for elements of interest were excluded.

TABLE 5.6 IRON SLAG STANDARD (FROM KRESTEN AND HJÄRTHNER-HOLDAR 2001, NORMALISED) COMPARED TO THE SEM RESULT (NORMALISED, AVERAGE OF THREE)

problem, three areas of each sample were analysed. A full table of results is shown in Table 5.4.

Analysis of Corning glass standards (Table 5.5) showed close agreement with their actual compositions. A slag standard was also analysed using the SEM, as its chemical composition is similar to the composition of the samples examined in this study. The SEM analysis of the slag sample agreed less well with its quoted composition and suggests that the iron, soda and alumina values determined by EDS may be high and the silica low (See Table 5.6 for the full analytical results and comments).

The bloom sample was etched using nital and then examined under a light microscope. Vickers hardness tests were also conducted on the microstructures within the bloom. A sample of the ore was analysed using X-ray diffraction (XRD).

RESULTS AND DISCUSSION

Slag

Two pieces of tap slag and one piece of smelting slag were analysed. The main elements present were iron and silicon. Of the other elements, there was a relatively high proportion of alumina with only a little lime and magnesia. The tap slag had less iron oxide present but more alumina (Al_2O_3) and silica than the smelting slag. The microstructure of the slag seen in Figure 5.12 was typical, containing predominantly iron oxide (wustite) and iron silicate (fayalite) in a glass matrix. Magnetite was occasionally noted.

Figure 5.13 compares the composition of the samples from The Butts with other slags from Worcester. Tube slag analysed from the Deansway excavation resembled the

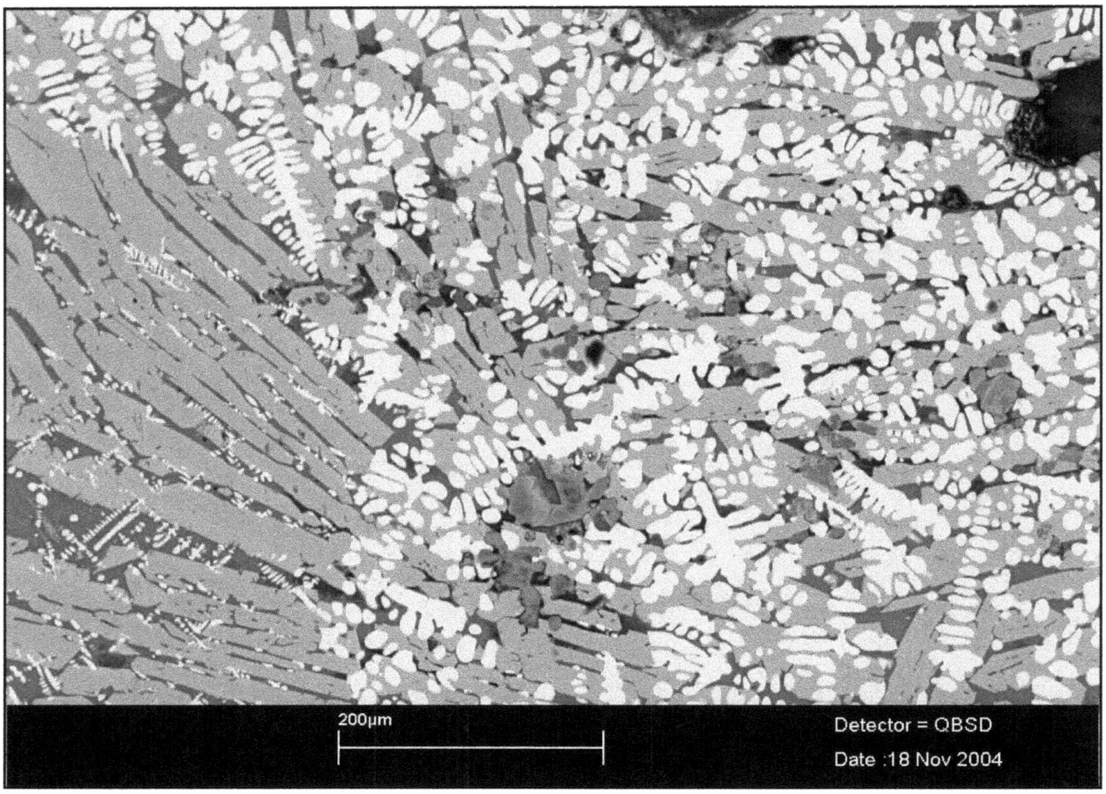

FIGURE 5.12 SEM BACKSCATTER IMAGE OF TAP SLAG FROM CONTEXT (4012) SHOWING THE MICROSTRUCTURE WHICH WAS PREDOMINATELY IRON OXIDE (WHITE) AND IRON SILICATE (LIGHT GREY) IN A GLASS MATRIX (DARK GREY)

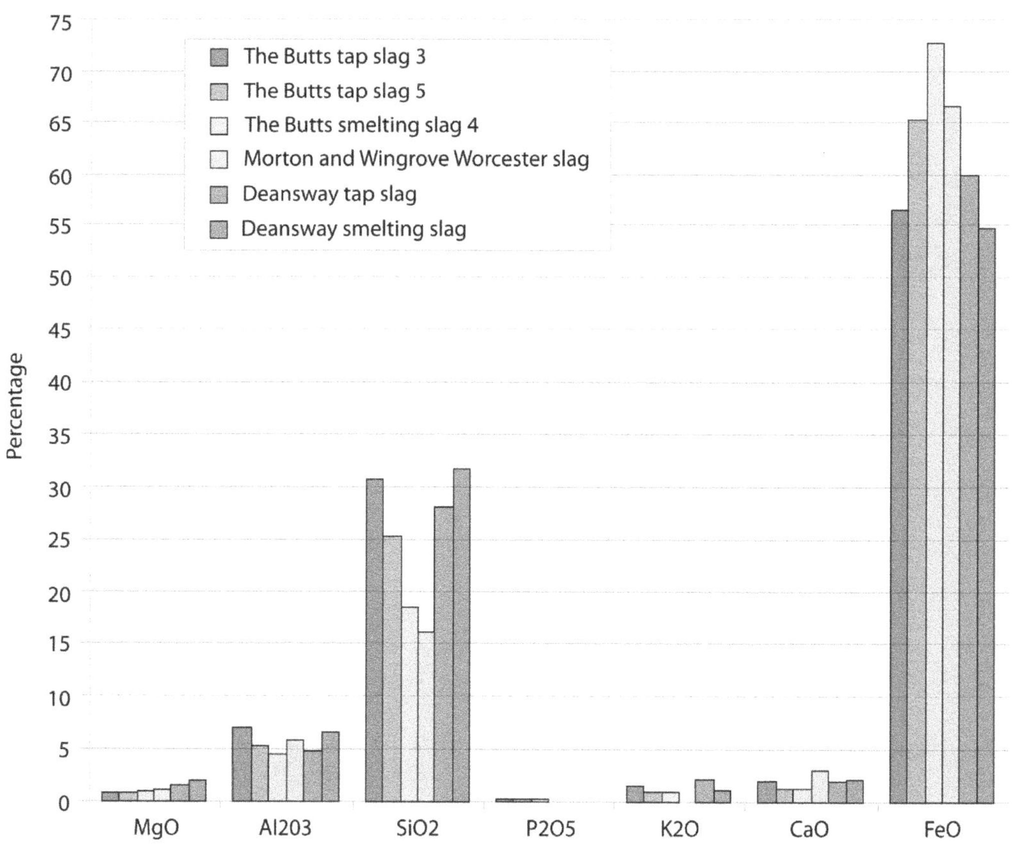

FIGURE 5.13 THE COMPOSITION OF SLAG FROM WORCESTER (DATA BASED ON AVERAGES FROM TABLE 5.5, AND FROM MORTON AND WINGROVE (1969) AND MCDONNELL AND SWISS (2004)

tap slag from The Butts, Worcester, both in composition and form (McDonnell and Swiss 2004). The Worcester slag analysed by Morton and Wingrove (1969, 1559) had high proportions of lime and alumina relative to the other gangue elements present in the tap slag. It is unknown where in Worcester this sample originated and whether it is a tap or smelting slag, although it is very similar in composition to the smelting slag found at The Butts, with a relatively large amount of iron oxide and low amount of silica.

The minor components within slags vary considerably across sites in different parts of the country. The composition is related to the raw materials, ore and clay lining, and the technological techniques, such as temperature, used during the smelting process. Compared to the concentration ranges in Paynter's study of regional variation in bloomery iron smelting, the Worcester slags have low concentrations of phosphorus and intermediate amounts of lime, magnesia, potash and alumina (Paynter 2006). This is typical for slags from the Gloucester, Worcester and Somerset area.

Ore

Geothite ores are predominant in the Carboniferous ore field of the Forest of Dean, which is thought to be the most likely source of iron ore for the smelting sites at Worcester (Fulford and Allen 1992). The chemical composition of these ores can vary considerably between areas within the Forest of Dean ore fields (Groves 1952, 179). Ore from this source could have been transported up the River Severn to the site at Worcester. Another possible source of ore is the fault-hosted ores of the Worcester Graben, in South Gloucestershire.

XRD showed that the sample of ore from The Butts was goethite with some haematite, and this was confirmed by SEM images which revealed small haematite crystals in a goethite matrix (Fig. 5.14). Tim Young (pers. comm.) has identified the ore as stalactitic and indicated that stalactitic ores are significant in the Forest of Dean, although stalactitic ores also occur in the larger voids elsewhere, for example the Glamorgan ore fields in South Wales. The ore found at Worcester is unlikely to have come from the Worcester Graben as these ores tend to be dominated by botryoidal facies rather than stalactitic facies (Young pers comm.; Young and Thomas 1999). Analysis of the ore showed that it contains 97% iron oxide and very small quantities of silica, alumina and magnesia. The ore found at The Butts was chemically similar to other goethite ores found at Deansway, Worcester (McDonnell and Swiss 2004). The low levels of lime and magnesia in the slag from The Butts are consistent with a high purity ore from the Forest of Dean being used during smelting.

Furnace lining

Analysis of the furnace lining from The Butts revealed that it was a quartz-rich clay containing over 70% silica. This

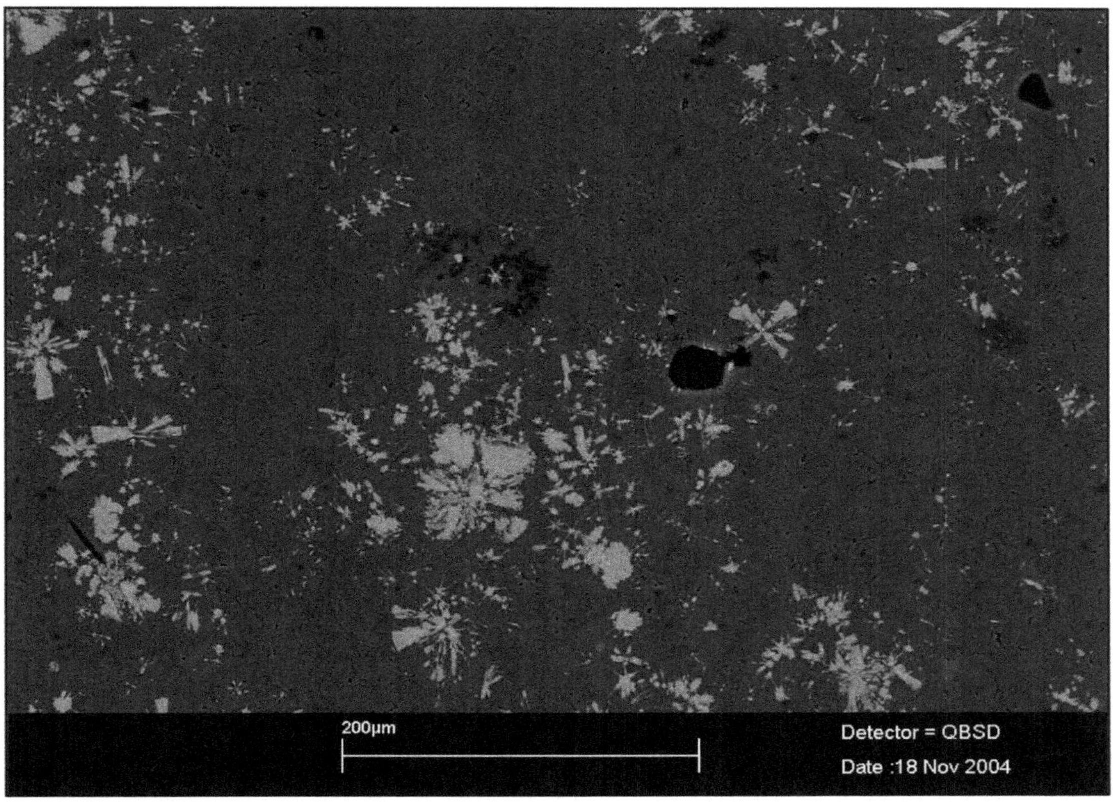

FIGURE 5.14 SEM BACKSCATTER IMAGE OF ORE FROM CONTEXT 1005 SHOWING THE SMALL LIGHT GREY HAEMATITE CRYSTALS IN A GREY GOETHITE MATRIX. THE BLACK INDICATES VOIDS WITHIN THE SAMPLE

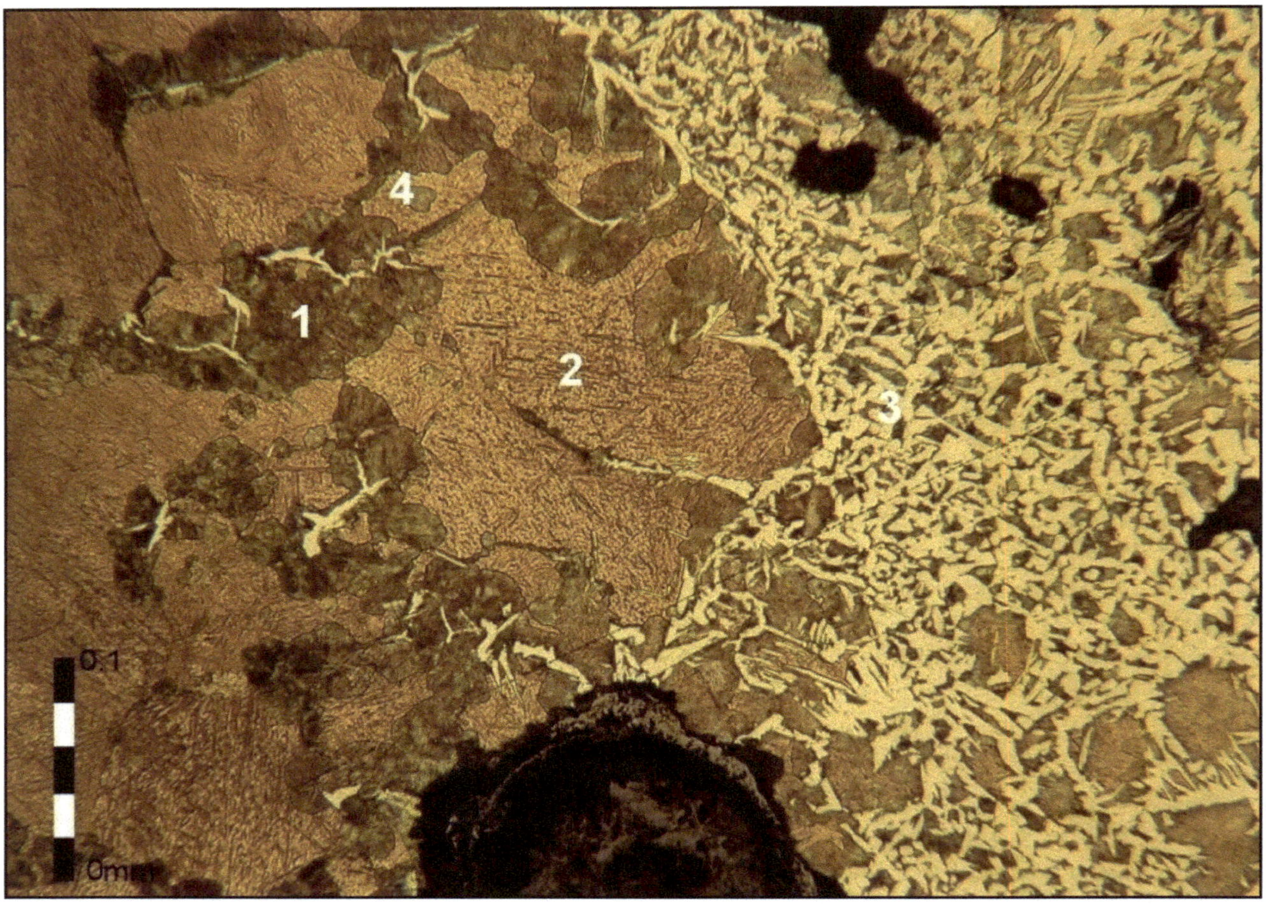

FIGURE 5.15 IMAGE OF THE BLOOM SHOWING: 1 – PEARLITE FORMING ON GRAIN BOUNDARIES; 2 – MARTENSITE; 3 – WIDMANSTÄTTEN FERRITE; 4 – BAINITE

appears to be a typical composition for Roman furnace linings and similar examples have been found during excavations at Westhawk Farm, Kent (Paynter 2002), Snettisham, Norfolk (Chirikure and Paynter 2004) and at the Chesters Villa, Woolaston (Fulford and Allen 1992, 192). The furnace lining also contained high amounts of alumina. Research has shown that the furnace lining may contribute a significant proportion to the overall composition of the slag (Fulford and Allen 1992). As the type of ore thought to have been smelted at the site contains little alumina (see above), the slightly high amounts of alumina present in the slag from The Butts may be derived from the furnace lining.

Bloom

The bloom was found in hollow way context 1005, which is a layer sealing a Roman cobbled surface. The bloom fragment weighed 520g and is 100mm by 60mm, and 60mm thick. No slag inclusions were found within it but analysis of slag surrounding the surviving metal revealed a similar composition to the slag found at the site, although the slag around the bloom had slightly lower amounts of alumina compared to the tap and smelting slag mentioned above.

Although hollow way layer 1005 was probably filled as part of post-medieval levelling activity, it contains a great deal of Roman material. The similarity between the chemical composition of the slag associated with the bloom and the smelting slag found in Roman contexts supports the conclusion that the bloom itself comprises redeposited Roman material.

The bloom was etched with nital to reveal the microstructure of the surviving metal and Vickers hardness tests were also conducted. The microstructure (Fig. 5.15) was heterogeneous and included phases of pearlite, bainite and martensite and a phase with Widmanstätten morphology, broad pointed needles (Samuels 1980, 313–15). The Widmanstätten phase is most likely to be ferrite and had a Vickers hardness value between 154 and 193. The proportions of these phases and the results of the hardness tests together indicate that the fragment is steel with a carbon content varying from 0.3% to 0.8%. The bloom fragment had also cooled unevenly; the presence of martensite in the central region indicating rapid cooling whereas the Widmanstätten microstructure of an adjacent region suggested slow cooling (*ibid*, 313–15). When the bloom fragment was analysed using the SEM the amount of phosphorus present throughout the sample was below the detection limit.

CONCLUSION

A significant quantity of ironworking waste of Roman date has been recovered from excavation at 14–24 The Butts. No furnace structures were found, but ironworking presumably occurred nearby. The similar date, chemical composition and form of the slag from The Butts, Farrier Street and Deansway suggest that they were produced using similar raw materials and processes, possibly at the same ironworking centre. This may have lain to the north of The Butts.

Both the microstructure and composition of the stalactitic ore suggest that the most likely source of the ore is the Forest of Dean. The microstructure of the bloom fragment from the site shows that it is a heterogeneous carbon steel with a carbon content of between 0.3% and 0.8%. The microstructure also indicates that the cooling rate varied across the bloom fragment. The pieces of bloom recovered from Deansway were all ferritic, so the bloom recovered from The Butts appears atypical. However, as this fragment of bloom was found discarded, it may not be representative of the iron being produced at the site.

ROMAN POTTERY *C Jane Evans, Nick Garland, Jon Millward, and Sian Reynolds with Jane Timby, D F Williams, and Felicity Wild*

SUMMARY

The bulk of the Roman pottery is grouped into two phases, both of which are based on the date of the ceramics rather than the site phasing detailed in the site narrative. The first section deals with the earlier Roman (ceramic Phase 1) material, which is discussed for each excavated area. The second section deals largely with the Roman assemblage recovered from the hollow way fill (1005), which is mostly 4th century (ceramic Phase 2), although this is residual material from a context dated to Phase 3 in the site narrative. The dating and functional composition of this assemblage is discussed in detail, including comparative discussions of vessel forms and fabrics. The other small group of late Roman pottery, from evaluation Trench 1, is considered after the section on the hollow way fill.

Table 5.7 shows that a total of 3,770 sherds of Roman pottery were recovered from the site, of which 3,420 sherds (weighing 65.8kg) came from the excavation, with a further 350 sherds from the evaluation and watching brief. A total of 1,846 Roman sherds were recovered from Phase 1 contexts which dated predominantly to the 2nd century, perhaps continuing into the early 3rd. These Phase 1 groups came from all four of the excavated areas (A, B, D and E), mostly from pits, ditches and gullies. Groups of Phase 2 late Roman pottery dating to the 4th century came from pits in Area E and evaluation Trenches 1 and 2, and from the fill of the hollow way in Area A (1005). This latter layer, however, also contained some medieval and post-medieval pottery and artefacts, which indicates that it was deposited during Phase 3 and therefore suggests that the ceramic Phase 2 pottery from layer 1005 is redeposited material. This sizeable late Roman pottery assemblage was, however, considered worthy of detailed analysis, as at the time, it was the only large group of later Roman pottery to have been recovered from Worcester. Other late Roman assemblages have occurred at 1 The Butts (Chapter 3) and 8–12 The Butts (Chapter 4) and subsequent excavation of the neighbouring New Library and History Centre site (in 2008 and 2009) has produced another large assemblage which, once analysed, will add to the data for late Roman Worcester presented here and in Chapters 3 and 4.

METHODOLOGY

The Area A assemblage, including the group from the hollow way fill (1005), was recorded and analysed by C Jane Evans. The remaining pottery was analysed by Nick Garland (Area B), Sian Reynolds (Area D) and Jon Millward (Area E), under the supervision of C Jane Evans. Full reports and the original data, in both digital

Area	Qty	% Qty	Wt (g)	% Wt	Average sherd wt (g)
A (1005)	1486	43.5	30546	46.4	21
A (other)	617	18.0	13670	20.8	22
B	345	10.1	2916	4.4	8
D	209	6.1	3332	5.1	16
E	675	19.7	13777	20.9	20
U/ S	88	2.6	1604	2.4	18
Excavation Total	*3420*	100.0	*65845*	100.1	19
Evaluation	323		9264		29
Watching brief/ well core	27		455		17
Total pottery	**3770**		**75564**		**20**

TABLE 5.7 SUMMARY OF ROMAN POTTERY BY AREA (INCLUDING RESIDUES)

WHEAS Fabric Code	ARCHIVE Fabric code	National Roman Fabric code: Tomber & Dore 1998	Common Name	Description/ references (T&D = Tomber and Dore 1998)
3	N02.1	MAL RE	Malvernian group A, handmade	T&D 147, plate 120; Peacock 1965–67
12	O02.1	SVW OX 2	Severn Valley ware	Standard oxidised fabric, unsourced: T&D 149, Pl 122; Webster 1976; Rawes 1982; Hurst and Rees 1992, 202
12.1	G04		Reduced variant	Standard fabric, reduced
12.2	O03.1		Organic variant, oxidized	Organic tempered variant, oxidised (elongated voids appearing as black/ dark grey streaks in fracture, Hurst and Rees 1992, 202
12.21	O03.11/O03.2		Fine organic variant, oxidized	Fine organic tempered, with sparse black organics, cf. Evans *et al.* 2000, 17 fabrics O1 and O5?
12.3	G05		Reduced organic variant	Darlington & Evans 1992, 37; Bryant and Evans 2004, 254–55
12.6	O02.2		Variant: yellow white inclusions	Bryant and Evans 2004, 256–57
12.5	O02.4		Variant: sandy, micaceous	Micaceous, with sparse, ill-sorted, rounded quartz <0.5mm, sparse black iron rich inclusions
13	O06.18		Sandy oxidised ware	Hurst and Rees 1992, 202
14	G012.3		Sandy reduced ware	Hurst and Rees 1992, 202
15	G012.1		Coarse sandy grey ware	Hurst and Rees 1992, 202–3
17	F018	PNK GT	Pink grog tempered ware	T&D 210, pl 175; Booth and Green 1989
19	N04	MAL RE A W	Malvernian A, wheelmade	Hurst and Rees 1992, 203
21.3	G012.5		Micaceous ware	Griffin 2002, 120; Griffin 2005a, 10
22	B02	DOR BB 1	Southeast Dorset BB1	T&D 127, pl 100; Williams 1977; Seager Smith and Davies 1993
23	H04	HAR SH	Harrold shelly ware	T&D 115, pl 89; Brown 1994
28	C01	LNV CC	Lower Nene Valley c. c. ware	T&D 118, pl 91 (Cream fabric)
28	C02	LNV CC	Lower Nene Valley c. c. ware	T&D 118, pl 91 (grey fabric)
28	C03	LNV CC	Lower Nene Valley c. c. ware	T&D 118, pl 91 (orange fabric)
29	C07	OXF RS	Oxfordshire red-slipped ware	T&D 176, pl 147
32	M02b	MAH WH	Mancetter Hartshill mortaria	T&D 189, pl 157a–d
33.1	M04c	OXF WH	Oxfordshire white mortaria	T&D 174, pl 145–46a–b
33.2	M04b	OXF WS	Oxford white-slipped mortaria	T&D 176, pl 147
33.3	M04a	OXF RS	Oxford red-slipped mortaria	T&D 176, pl 147
34?	M07?		West Midlands mortarium	Jane Timby notes 'Orange buff exterior, pink core with white margins. A hard, granular, sandy fabric, possibly with quartz trituration grits. The fabric is one known from the West Midlands area and is known to occur in Warwickshire (P. Booth pers. comm.) but is of unknown provenance.' Possibly related to WHEAS fabric 34, thought to come from Wroxeter
38	P09	OXF WH	Oxford White ware	T&D 174, pl 145–46a–b
40	P08	OXF PA	Oxford parchment ware	T&D 174, pl 145
42.1	A02	BAT AM 2	Amphorae, Baetican Dressel 20	Peacock and Williams 1986, class 25; T&D 85, pl 62
43	S02	LMV SA	Les Martres-de-Veyre samian	T&D 30, pl 19
43	S02/ 3		Central Gaulish Samian	See Wild below
44	C04	CNG CC 2	Central Gaulish c. c. ware 2	T&D 53, pl. 38; Symonds 1992, 63–64
114	D03		Mica dusted ware	Bryant and Evans 2004, 264
115	C011	NFO CC	New Forest (metallic) c.c. ware	T&D 141, pl 115
149	G012.2		Worcestershire imitation BB1	Griffin 2005a, 13
151	C012		Micaceous colour coated ware	Wheelmade, hard, micaceous fabric with abundant silt sized quartz. Fired reddish yellow (7.5YR 6/ 6) with a red colour coat (2.5YR 5/ 6)

TABLE 5.8 LIST OF FABRICS REPRESENTED

and hard copies, are included in the archive. Summaries of the pottery from the evaluation and watching brief are also included below, although the pottery was not recorded in detail. Nor was the pottery recovered from the soil residues analysed in detail; but is included in the overall quantification in Table 5.7.

All pottery was recorded using the Birmingham Archaeology pottery recording system, described in the archive. Fabrics were cross referenced with the Worcestershire County Fabric Series, prefixed WHEAS in the text (Hurst and Rees 1992, 200–209; <http://www.worcestershireceramics.org>) and, where possible, the National Roman Fabric Reference Collection (Tomber and Dore 1998). Sherds were examined by eye and using a x10 hand lens, or occasionally a x20 binocular microscope. The assemblage was quantified by sherd count, weight and rim EVE (estimated vessel equivalent); base EVEs are recorded in the archive but not published. Precise form types and broad vessel classes (eg, bowl, cook pot) were recorded, together with any evidence for decoration, manufacture, repair, use or reuse. Data analysis was undertaken using Microsoft Access 2002 and Excel 2002. This digital data is included in the archive.

FABRICS

The fabrics represented in the pottery assemblage from 14–24 The Butts are listed in Table 5.8.

Ceramic Phase 1: 2nd- and 3rd-century assemblages

Introduction

The pottery assigned to this phase was recorded in detail and was retrieved from the following features: a ditch at the southern extent of Area A; pits in Area B and Trench 3; a cobbled surface and pits in Area D; pits and gullies in Area E and Trench 2; evaluation Trench 4; the cut and fill of a well (F901) in watching brief Plot 5.

Based on groups recovered from all these features, the earliest activity on the site appears to date to the early 2nd century. Evidence for this comes from a handful of sherds of Les-Martres-de-Veyre samian ware, including a decorated form 37 bowl dated to *c* 100–120 from a small gully in Area E (4013, Wild, this chapter). An early 2nd-century start date is supported by the coin evidence (Nogueira, this chapter), and would also be contemporary with activity noted on the nearby site at 8–12 The Butts (Chapter 4, this volume).

Table 5.9 and Figure 5.16 show that there is some variation in the fabric profiles for these predominantly earlier groups. All, however, are dominated by Severn Valley ware, particularly the characteristically earlier organic-tempered variant. A number of the coarse ware forms date broadly to the late 1st to early 2nd century; these are described in the summaries below. There is no evidence to support definitively 1st-century activity, however, and the absence of characteristically 1st-century South Gaulish samian ware is noted in the samian ware report (Wild, this chapter). The bulk of this assemblage, therefore, appears to relate to 2nd-century activity. The coarse wares include diagnostically 2nd-century forms, as well as forms more broadly datable to the 2nd to 3rd century. However, the latest samian, a form 45 mortarium, dates to *c* 170–200 (Wild, this chapter). This is broadly contemporary with the coins of Septimius Severus, dating to 193–211 and 195 and discussed in the small finds report (Nogueira, this chapter).

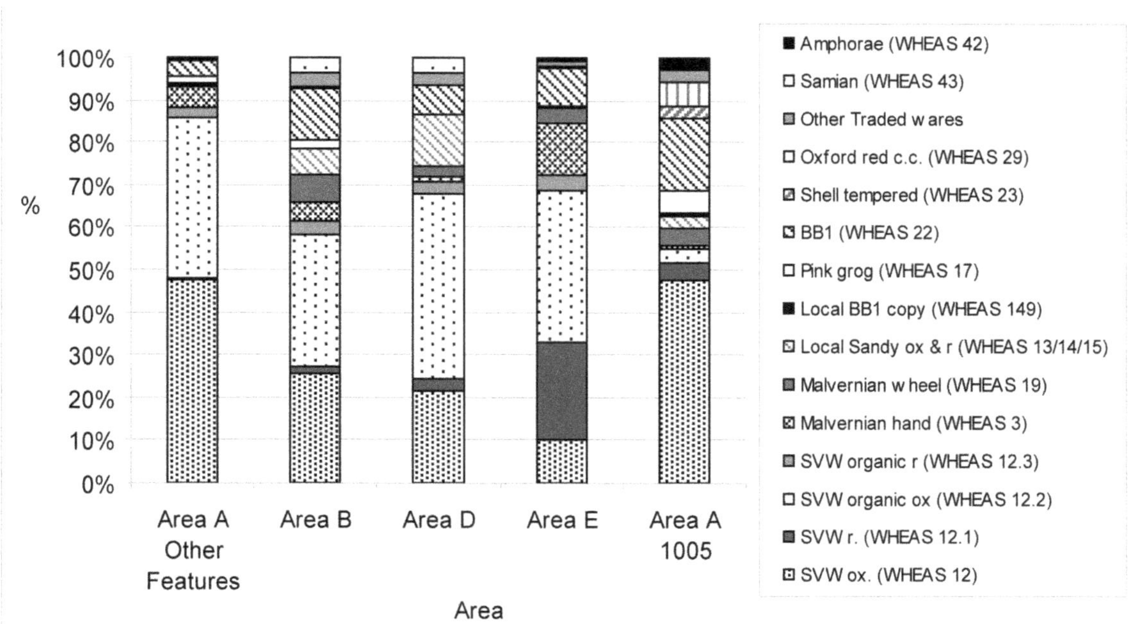

FIGURE 5.16 SUMMARY OF NON-KEY GROUP ASSEMBLAGES BY FABRIC (% WEIGHT)

WHEAS Fabric Code	Area A Qty	Wt (g)	Area B Qty	Wt (g)	Area D Qty	Wt (g)	Area E Qty	Wt (g)
3	52	661	13	122	2	41	107	1657
12	301	5941	107	751	52	715	81	1380
12.1	8	90	5	43	5	92	108	3088
12.2	128	5076	0	0	0	0	16	650
12.21	0	0	92	904	72	1455	112	4209
12.3	12	348	12	97	7	92	13	455
12.6	18	469	0	0	0	0	0	0
13	2	65	3	20	4	111	0	0
14	1	4	4	32	21	305	2	28
15	1	3	6	104	0	0	0	0
19	4	49	17	190	7	73	12	512
149	0	0	0	0	0	0	0	0
21	0	0	0	0	0	0	0	0
114	1	6	0	0	0	0	0	0
34?	0	0	1	111	0	0	0	0
Regional wares	*528*	*12712*	*260*	*2374*	*170*	*2884*	*451*	*11979*
22	35	476	64	357	22	223	74	1196
28	5	25	1	2	0	0	0	0
29	0	0	7	17	0	0	2	31
17	2	183	1	61	0	0	1	73
23	0	0	0	0	1	9	0	0
32	0	0	0	0	1	14	0	0
33.3	0	0	0	0	0	0	0	0
33.2	0	0	0	0	0	0	0	0
33.1	3	69	0	0	0	0	0	0
40	0	0	6	49	0	0	2	27
38	1	1	0	0	0	0	0	0
115	0	0	0	0	0	0	0	0
151.1	0	0	0	0	0	0	0	0
Traded wares	*46*	*754*	*79*	*486*	*24*	*246*	*79*	*1327*
43.2	3	42	*5*	54	1	120	4	40
42.1	0	0	*0*	0	0	0	2	65
44	0	0	*1*	2	0	0	0	0
Imported wares	*3*	*42*	*6*	*56*	*1*	*120*	*6*	*105*
Misc. White c.c.	0	0	*0*	0	8	60	9	166
Misc. White	0	0	*0*	0	4	16	0	0
Misc. c.c.	0	0	*0*	0	0	0	1	4
Unknown	*0*	*0*	*0*	*0*	*12*	*76*	*10*	*170*
Total	**577**	**13508**	**345**	**2916**	**207**	**3326**	**546**	**13581**

TABLE 5.9 SUMMARY OF NON-KEY GROUP ASSEMBLAGES BY FABRIC (EXCLUDING RESIDUES)

It seems likely, therefore, that the main period of deposition finished around the very end of the 2nd century, or very early in the 3rd century. For subsequent periods, a number of broadly 3rd- to 4th-century forms were noted, but nothing that need necessarily pre-date the late 3rd century; there was no East Gaulish samian ware for example. Diagnostically late 3rd- and 4th-century wares were noted in some features, for example Oxfordshire red colour-coated ware and pink grog-tempered ware. The other 3rd- to 4th-century forms could therefore be contemporary with these and with the Area A hollow way fill deposit discussed separately below.

Area A and Trench 1

A total of twelve contexts, from six features, in Area A produced 617 sherds of pottery, weighing 13.7kg, with a high average sherd wt (22g). The pottery came from the

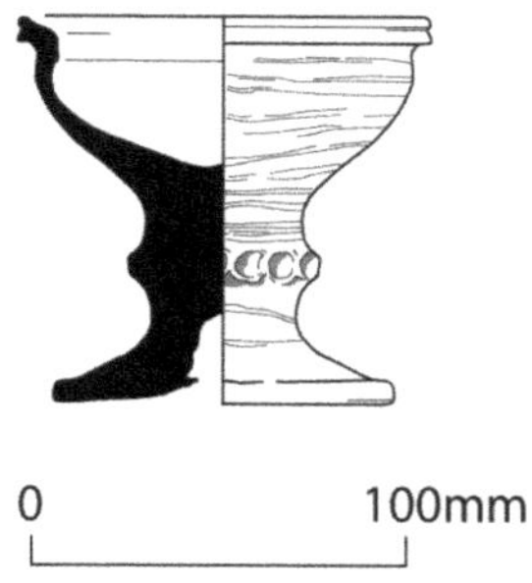

FIGURE 5.17 *TAZZA* FROM PIT F106

following features and contexts (see Figs 5.2 and 5.3): F100 (1011, 1012); F106 (1016); F107 (1015, 1023); F108 (1018); F109 (1020, 1028); and from contexts 1003, 1004, 1014 and 1017). The majority (68% by count and 82% by weight) came from a Phase 1 ditch (F109/ F107), mainly from its upper fill (1015, 1020) and a layer of slumping on its northern edge (1017). These fills produced a Central Gaulish samian ware beaker rim (Wild, this chapter) dating to the 2nd century, and a range of Severn Valley ware, Malvernian and Dorset Black Burnished ware (BB1) forms broadly consistent with this date (cf. Webster 1976, fig. 1, A4; fig. 4, C20; fig. 7, E38–43; fig. 9, H59, 60; Peacock 1965–67, fig. 1, 1–10; Seager Smith and Davies 1993, fig. 122, type 1). A couple of forms (cf. Webster 1976, fig. 1, A7; Evans *et al* 2000 fig. 24, type 5, JWM15) were more typical of the 3rd to 4th century, and may be intrusive. This ditch was cut by a small pit (F106, 1016) which contained an almost complete *tazza*, presumed to date to the 2nd century and illustrated in Figure 5.17 as well as on the front cover. As incense burners, *tazze* have often been classified as 'ritual' vessels, with religious associations. However, the large number recently recovered from a Roman domestic site at Gresham Street (GSM97), London, suggests that they may also have functioned as more everyday lamps or incense burners (Featherby 2004).

No pottery was recovered from the cobble surface (1006, Phase 2), but the overlying layer (1012) included a late 3rd- to 4th-century BB1 jar (Seager Smith and Davies 1993, 231, fig. 122, type 3) and a flange-rimmed bowl or dish (*ibid*, fig. 124, type 25), as well as 2nd-century types. Small quantities of contemporary material were also recovered from Trench 1, including samian ware (see Wild, this chapter).

As can be seen in Figure 5.16, the assemblage as a whole was heavily dominated by Severn Valley ware, and included a range of diagnostically 1st- to 2nd-century fabrics, namely handmade Malvernian ware, organic-tempered Severn Valley ware and a sherd of mica dusted ware (WHEAS fabrics 3, 12.2, 12.3, 114). The latter was widespread in Britain in the early 2nd century (Marsh 1978, 122). Decoration, where datable, was typical of the 1st to 2nd century (ie, linear rustication on jars) and 2nd century (ie, acute lattice burnishing on BB1 jars). However, occasional sherds in pink grog-tempered ware (WHEAS fabric 17, contexts 1011, 1012) provided further evidence for intrusive, later material.

Area B and Trench 3

One of the smallest and most fragmentary assemblages, with the lowest average sherd weight (Table 5.7), came from Area B. Most of the pottery (255 sherds) came from a series of Roman pits dated to the 2nd and early 3rd centuries. The most substantial group (186 sherds) came from the large cut at the southern limit of Area B (F200, 2004, see Figs 5.4 and 5.5). The other Roman pits (F202, F203, F204, F205 and F208) produced smaller quantities, all broadly dating to the 2nd to 3rd century. A total of 34 sherds were residual in post-medieval features (F210, F213, F216, F218, Phases 3 and 4).

The earliest pottery dated broadly from the late 1st to early 2nd century, including a Severn Valley ware carinated bowl (Webster 1976, fig. 9, H59, 60), and a rusticated jar. These are broadly contemporary with the Trajanic coin, issued between AD 101 and 102, found in a pit F207 (Nogueira, this chapter). Other forms, Malvernian tubby cooking pots (Peacock 1965–67, fig. 1, 10 and fig. 1, 5, 7–9) and a Severn Valley ware jar (Webster 1976, fig. 4, C20) could also be contemporary, although they continued in use longer into the 2nd century. A coin dated to AD 193–211 from F200 supports a late 2nd- or early 3rd-century date for these features (Nogueira, this chapter). The samian ware forms also dated to the latter half of the 2nd century: a form 31 bowl from F200, a form 79 dish from F203 and a scrap of Antonine samian ware from F210 (Wild, this chapter). Various coarse ware forms are consistent with this dating: a 2nd-century BB1 jar (Seager Smith and Davies 1993, fig. 122) and BB1 and Severn Valley ware forms dated more broadly to the 2nd to 3rd century (*ibid*, fig. 122; Webster 1976, fig. 1, A3; fig. 4, C22; fig. 7, E40–3).

There was also some later Roman pottery found in this area; a typically 3rd-century Severn Valley ware jar (Webster 1976, fig. 3, A8) and a late 3rd- to 4th-century BB1 jar (Seager Smith and Davies 1993, fig. 122). Figure 5.16 shows that the fabric profile of this group is more complex than that for the Area A assemblage described above, perhaps reflecting the greater chronological range of material incorporated in this assemblage.

The date of activity associated with Trench 3 was hard to define as most of the pottery comprised undiagnostic sherds of Severn Valley ware. One of the pits (F305E) produced a hooked-rim jar and a flange-rimmed bowl, both broadly 2nd- to 3rd-century types. This feature and pit F303E also produced Malvernian copies of BB1 plain-rimmed dishes, a type produced from the 2nd century onwards. Only levelling layer 3009E produced BB1; two undecorated body sherds from a cook pot, and a single sherd of samian ware.

Area D

There were no coins or other small finds to provide dating in this area. The pottery assemblage is also small, but with a fairly high average sherd weight (Table 5.7). The pottery studied came mainly from two Phase 1 features (Fig. 5.6): pit F305 (fill 3006) produced only eight sherds, most of the assemblage coming from the east–west ditch truncating this (F304, 3004, 3005). This assemblage produced a relatively high proportion of samian ware. The latest piece was a central Gaulish, form 45 mortarium dated *c* AD 170–200, from the upper fill of the ditch (3004, Wild, this chapter). However, this context also produced a sherd of late Roman shell-tempered ware, suggesting that some intrusive, later Roman material was included in the assemblage. Earlier samian ware included a Hadrianic decorated sherd from the same context (Wild, this chapter, no. 2), and sherds of Hadrianic to early Antonine samian ware from cobble layer F300 (3001).

The proportions of Severn Valley ware and BB1 are very similar to those for Area B (Fig. 5.16), though with a higher proportion of organic-tempered Severn Valley ware. The coarse ware forms and fabrics are consistent with a broadly 2nd-century date. BB1 forms included a type 22 flat-rimmed bowl (Seager Smith and Davies 1993, fig. 123; Reynolds 2005, fig. 13, 27), and a type 1 cook pot (Seager Smith and Davies 1993, fig. 122; Reynolds 2005, fig.13, 24), both typical 2nd-century types. Organic-tempered Severn Valley ware was again the most common variant (Fig. 5.16). Forms included wide-mouthed jars with triangular rims (Reynolds 2005, fig. 8, 5, 9; Webster 1976, fig. 4, C22) and a tankard with moderately splayed walls (Reynolds 2005, fig. 8, 6; Webster 1976, fig. 7, E41), all broadly dated to the 2nd to 3rd century. Also present were wide-mouthed jars with flanged rims, similar to forms included in the late Roman hollow way deposit (1005, Fig. 5.16). These are associated with late Roman deposits in Area A, and some other sites (Peacock 1965–67, fig. 3, 32–51), but similar forms are published from 2nd- and early 3rd-century deposits elsewhere (Rawes 1982, fig. 4, 68, 69; Evans *et al* 2000, fig. 24, JWM15).

Area E and Trench 2

The Area E assemblage was again dominated by Severn Valley ware (Fig. 5.16). The proportion of organic-tempered Severn Valley ware was similar to the assemblages described above, but this assemblage produced a significantly higher proportion of reduced Severn Valley ware (Fig. 5.16). Most pottery came from pits (Figs 5.7 and 5.8), with eleven residual sherds coming from Phase 3 ditches (F415, F417, F418) and 122 from post-medieval layers (4003, 4034, 4035). This was one of the least fragmentary of the assemblages; it had the highest average sherd weight (Table 5.7, 20g), produced a number of complete profiles and had a number of joining sherds.

The largest single group came from a possible gully terminal (F405, 4027, 4013) which appears to have been used as a rubbish dump. The upper fill (4013) produced 98 sherds, with a particularly high average weight of 37g. The lower fill (4027) produced 123 sherds, 99 of which came from a near complete Malvernian tubby cooking pot; this explains the relatively high proportion of this ware in the group (Fig. 5.16). The samian ware from this feature, all from the upper fill, comprised a stamped base, dated *c* 125–50, a decorated sherd dated *c* 100–120 (Wild, this chapter, no. 1) and a Hadrianic or Antonine bowl, with a rivet hole indicating repair. The coarse wares included types dating to the late 1st or early 2nd century, for example a rusticated jar (Millward 2005, 1032), and a range of Severn Valley ware forms broadly dating to the 2nd to 3rd century (Webster 1976, fig. 5, C23, fig. 7, E40; Evans *et al* 2000, fig. 26, type 2 bowl).

The samian ware from the rubbish pit truncating F405 (F404, 4012, 4015, 4022) dated to the second half of the 2nd century. This feature also contained a 2nd-century BB1 flat-rimmed bowl (Seager Smith and Davies 1993, fig. 123, type 22). The pottery from the other pits supported a 2nd- or early 3rd-century date, with pit F402 producing a 2nd- or early 3rd-century tankard (Webster 1976, fig. 7, E40).

As Figure 5.7 indicates, many of the ditches in Area E date to the 17th and 18th centuries, with ditch (F417) producing two residual sherds, both in the characteristically early organic-tempered Severn Valley ware. Pit 406 (which cut F417) produced a coin (Nogueira, this chapter), issued in AD 195. Although this was residual in a Phase 3 feature, it is indicative of late 2nd-century activity in this area.

Some of the latest Roman forms came from the upper fill of a large Phase 2 pit (F400, 4005). These included Oxfordshire colour-coated ware (WHEAS fabric 29) and pink grog-tempered ware (WHEAS fabric 17) and a residual sherd of Antonine samian ware. Further late Roman residual pottery came from a layer (4003, 17th to 18th century), and from the overburden (4035). The former included a jar similar to types found in Area A (1005; Webster 1976, fig. 7, C32; Peacock 1965–67, fig. 3, 52–54, fig. 4, 55–60). The latter included a type 25 flanged BB1 bowl (Seager Smith and Davies 1993, fig. 124), a Severn Valley ware bowl (cf. Evans *et al* 2000, 37, fig. 27, BT.13; Peacock 1965–67, fig. 3, 35) and a Malvernian storage jar (Peacock 1965–67, fig. 4. 88–89), all typically late forms.

The pottery from Trench 2 was not particularly well preserved. The diagnostic fabrics and forms indicated a broadly 2nd-century date including organic-tempered Severn Valley ware (WHEAS fabric 12.2, 12.3), reduced Severn Valley ware (WHEAS fabric 12.1) and Mancetter-Hartshill mortaria (WHEAS fabric 32). The latter had quartz and sandstone trituration grits typical of earlier vessels. All but ditch F200E produced BB1, indicating a *terminus post quem* of *c* 120, although small quantities were reaching the area earlier than this. Pit F202E, however, also produced a coin dated 270+ (Nogueira, this chapter), suggesting that the pottery here at least is likely to be residual. This pit also produced some unusual Malvernian tile, which is discussed further elsewhere (Macey-Bracken, this chapter).

Evaluation Trench 4

Of most interest was the assemblage from the earliest deposit (4007E) in Trench 4 (Fig. 5.1). This produced substantial, unabraded sherds, many of which were in chronologically diagnostic forms. Most sherds dated broadly to the 1st or early 2nd century, although some slightly later material was also present. The evidence from the excavated assemblages discussed above suggests that a date in the early 2nd century is most likely. Severn Valley ware was the most common fabric, usually organic-tempered (WHEAS fabric 12.2, 12.3). Forms included an upright walled tankard (Webster 1976, fig. 7, 38), narrow and wide-mouthed jars (*ibid*, fig. 1.3, fig. 4.20), and more unusual forms such as a bead rimmed dish and a carinated bowl with a flaring rim. One of the narrow-mouthed jars had a *graffito* on the rim (Tomlin, this chapter). Severn Valley ware jars with similar graffiti are published from Alcester (Evans *et al* 1994, 124–30) and other examples are known from *Ariconium* (Willis 2000) and Wroxeter (Evans 2000). The latest form was a broadly mid 2nd- to late 3rd-century type (Webster 1976, fig. 5, 23), in plain Severn Valley ware (WHEAS 12). Forms in reduced Severn Valley ware (WHEAS 12.1) included a jar with linear rustication and a flask, again consistent with a 1st- to early 2nd-century date. Other fabrics included handmade Malvernian ware (WHEAS fabric 3), represented by a very large storage jar (Peacock 1965–67, fig. 4, 80) and a lid, and BB1 (WHEAS fabric 22), represented by a jar with a short, slightly everted rim. While the presence of BB1 generally indicates a *terminus post quem* of *c* 120, small quantities were reaching this area earlier than this, and the jar is a type found most commonly at Dorchester, Greyhound Yard in contexts dated 75–120 (Seager Smith and Davies 1993, fig. 122, type 1). A single sherd of Dressel 20 amphorae was recovered, but no samian ware.

Watching brief: Plot 5, well F901

Excavation in Plot 5 (Fig. 5.1) produced 20 sherds weighing 371g, providing some useful dating evidence. The fill of the cut for the well (9000) produced five sherds, amongst which was a rim from a 3rd-century Severn Valley ware jar (Webster 1976, fig. 3.8), a sherd from a BB1 jar decorated with right angle cross hatch, dating to *c* 160–280 and a sherd of handmade Malvernian ware, probably dating to the 2nd century. The upper fill (9002), thought to represent a single episode of back-filling once the well was abandoned, produced twelve sherds. These included the splayed rim from a BB1 cook pot decorated with obtuse cross hatch burnish, dating to the late 3rd or 4th century (Seager Smith and Davies 1993, fig. 122, type 3; cf. Fig. 5.20, 52). The well core produced only three sherds. From 1–2m down came a Severn Valley ware body sherd with a handle scar, probably from a flagon. From 3–4m down came a single, small body sherd in a micaceous fabric with dark surfaces. This is probably WHEAS fabric 21.3, first identified on the nearby New Police Station and Magistrates's Court sites on Castle Street. If so, it dates to the 1st to 2nd century.

None of the other sherds from the watching brief, 2 from Plot 1 and 5 from Plot 3 were diagnostic.

Ceramic Phase 2: The late Roman assemblages

Introduction

The cobbled stone surface F100, (1006), contained within an east–west hollow way and interpreted as a late Roman thoroughfare, was covered by a garden-like soil which filled the hollow way, and contained significant quantities of Roman finds, mixed with medieval and post-medieval material. The presence of medieval and post-medieval material suggests that the filling of the hollow way took place in the late medieval or post-medieval (site Phase 3) period. With scant evidence for activity between the late Roman and late medieval periods, and despite comprising redeposited material, the Roman assemblage from layer 1005 can be considered as a chronologically distinct group. The other small group of late Roman pottery, from evaluation Trench 1, is considered after the section on the hollow way fill.

Hollow way fill (1005)

The fill of the hollow way (1005) produced a total of 1,463 Roman sherds (Table 5.7), weighing 30,446g, with a total rim EVE of 21.These included diagnostically late Roman fabrics (Table 5.10) and forms (Figs 5.18–5.20), providing good dating evidence for the assemblage. The relative proportions of diagnostically late Roman forms are presented graphically in Figure 5.21; these forms are described in the illustration catalogue below (Figs 5.18–20). Most of the pottery is of 4th-century date. Earlier (ceramic Phase 1) material comprises three sherds of Hadrianic–Antonine or Antonine samian ware (Wild, this chapter), two of Dressel 20 amphorae (Williams, this chapter), and a couple of typically 2nd-century BB1 forms (Seager Smith and Davies 1993, fig. 122, type 1; fig. 123, type 22). Fabrics such as organic-tempered Severn Valley ware (WHEAS fabric 12.2) and handmade Malvernian ware (WHEAS fabric 3), together with a few other, long lived fabrics and forms, are also almost certainly earlier, though the Malvernian ware forms were not datable types.

Shell-tempered ware

Of particular note is the presence of wheelmade, shell-tempered ware (WHEAS Fabric 23) which dates to the later

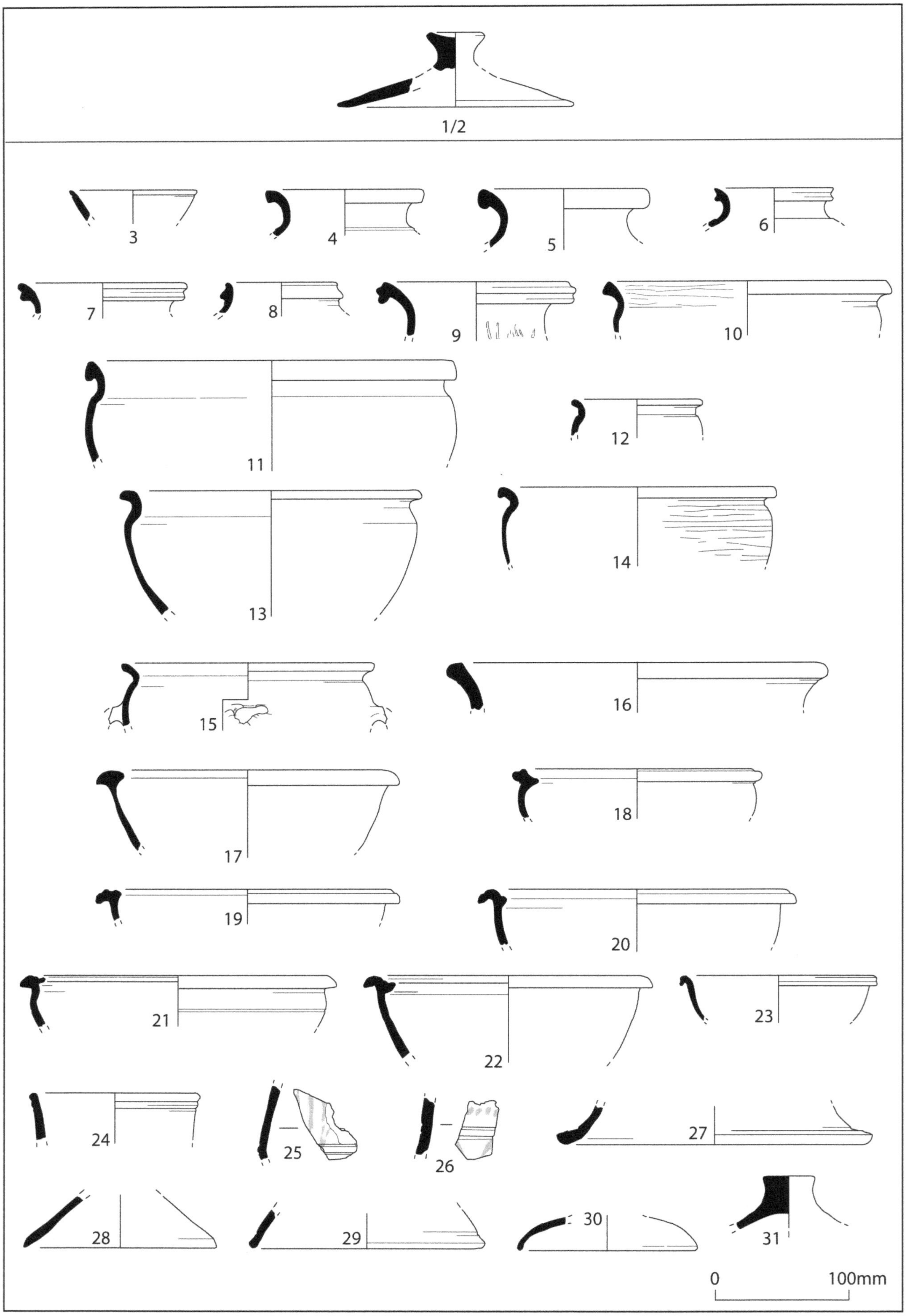

Figure 5.18 Pottery, catalogue forms 1–31

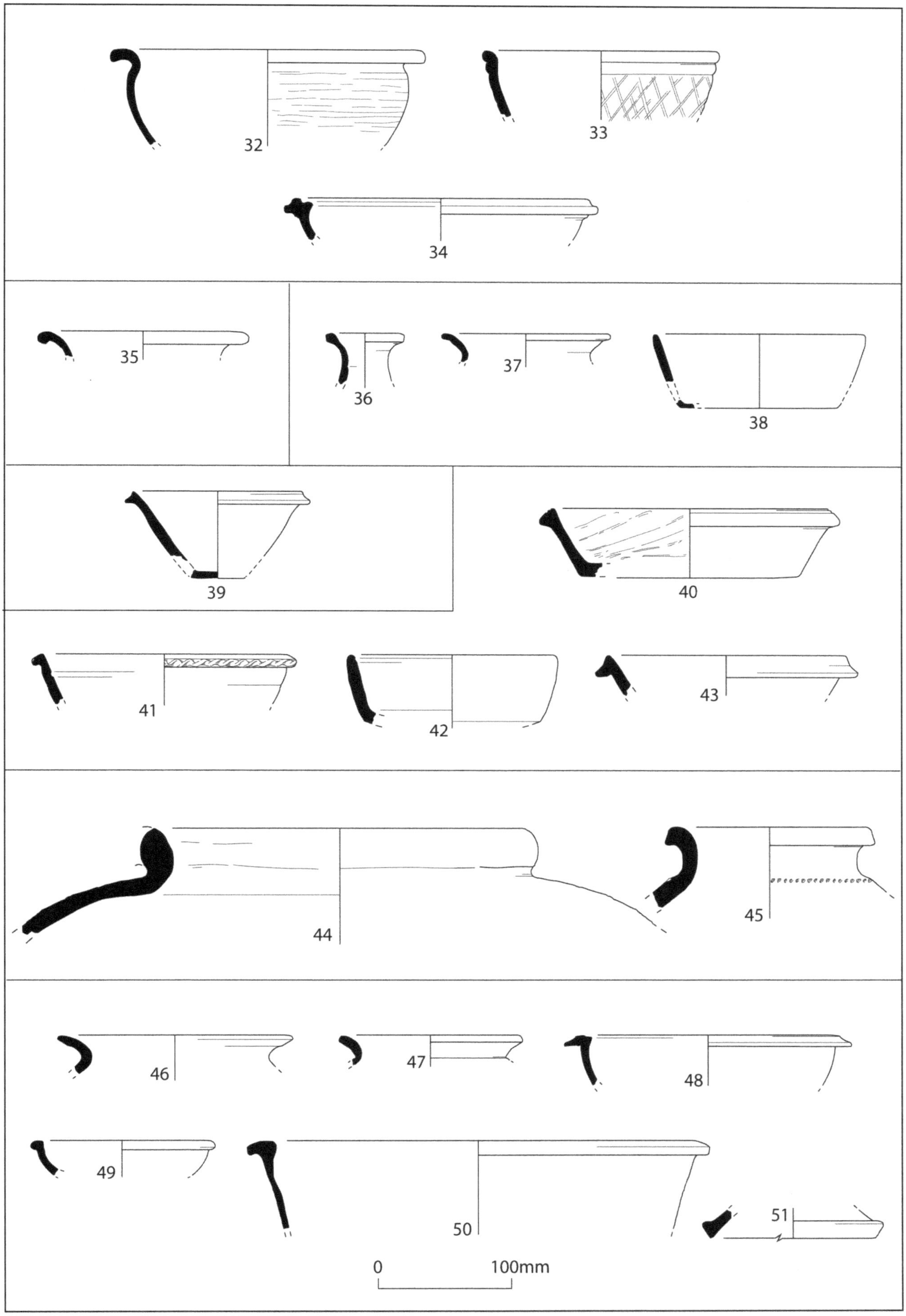

Figure 5.19 Pottery, catalogue forms 32–51

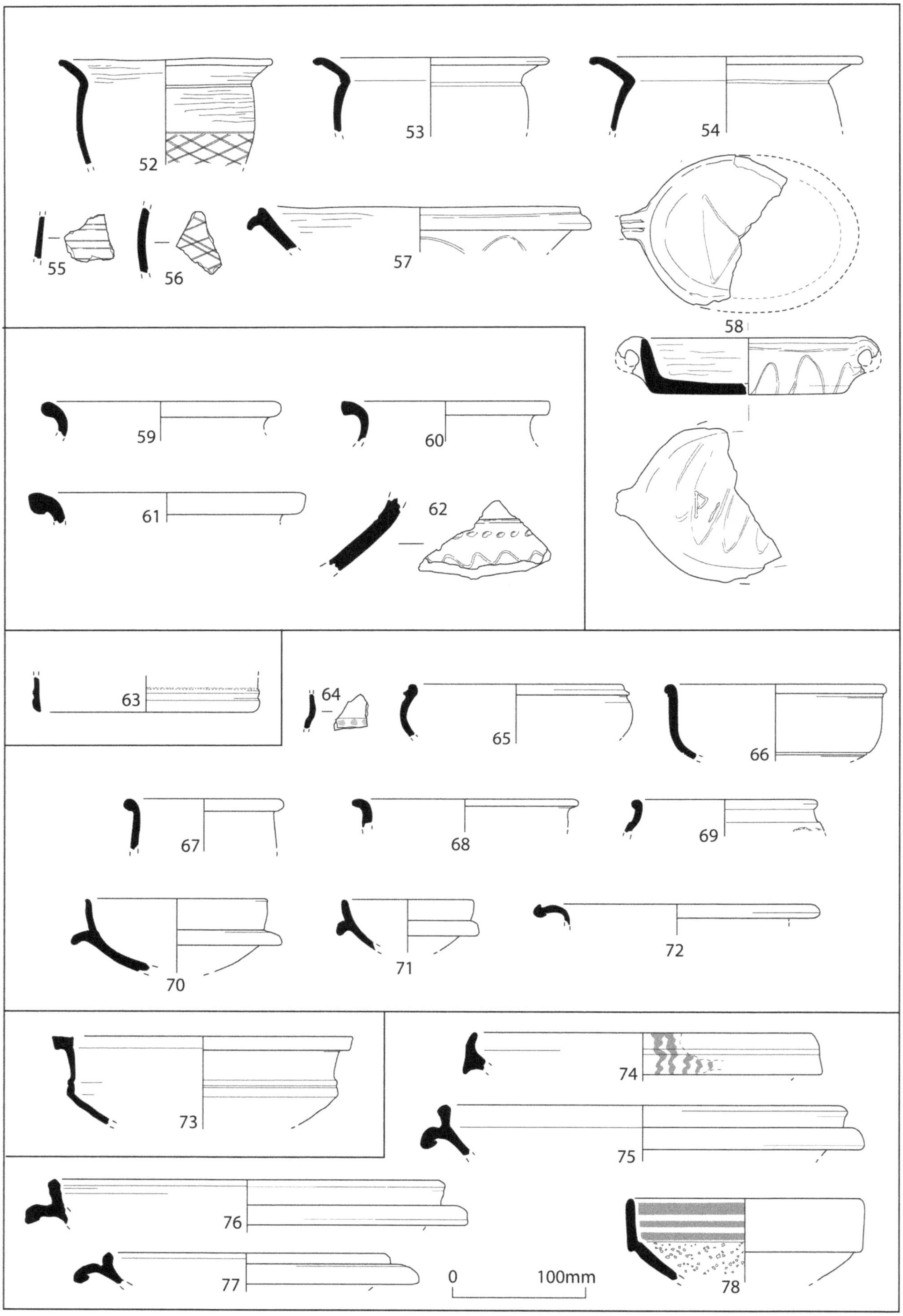

FIGURE 5.20 POTTERY, CATALOGUE FORMS 52–78

WHEAS Fabric Code	Qty	% Qty	Wt (g)	% Wt	% Rim	% Rim EVE	Average wt (g)
3	9	0.6	227	0.7	9	0.4	25
12	640	43.9	10647	35.0	926	41.2	17
12.1	39	2.7	995	3.3	29	1.3	26
12.2	18	1.2	283	0.9	26	1.2	16
12.21	32	2.2	549	1.8	0	0.0	17
12.3	1	0.1	3	0.0	0	0.0	3
12.6	39	2.7	538	1.8	40	1.8	14
12?	47	3.2	599	2.0	8	0.4	13
13	5	0.3	67	0.2	23	1.0	13
14	3	0.2	25	0.1	9	0.4	8
14.2	3	0.2	14	0.0	8	0.4	5
15	21	1.4	624	2.0	56	2.5	30
19	40	2.7	980	3.2	145	6.4	25
149	13	0.9	171	0.6	39	1.7	13
21	3	0.2	36	0.1	0	0.0	12
Regional	*913*	*62.6*	*15758*	*51.8*	*1318*	*58.6*	*17*
22	271	18.6	4243	13.9	591	26.3	16
Misc. c.c	4	0.3	36	0.1	0	0.0	9
28	11	0.8	121	0.4	0	0.0	11
28	10	0.7	72	0.2	14	0.6	7
28	14	1.0	143	0.5	26	1.2	10
29	116	8.0	1402	4.6	182	8.1	12
115	5	0.3	22	0.1	0	0.0	4
151	1	0.1	4	0.0	0	0.0	4
17	17	1.2	1305	4.3	29	1.3	77
23	23	1.6	670	2.2	21	0.9	29
32	12	0.8	4568	15.0	7	0.3	381
33.3	8	0.5	164	0.5	15	0.7	21
33.2	1	0.1	45	0.1	8	0.4	45
33.1	18	1.2	115	0.4	17	0.8	6
40	19	1.3	213	0.7	16	0.7	11
38	2	0.1	60	0.2	0	0.0	30
Traded	*532*	*36.5*	*13183*	*43.3*	*926*	*41.2*	*25*
44	2	0.1	3	0.0	0	0.0	2
43.2	12	0.8	89	0.3	6	0.3	7
42.1	4	0.3	1413	4.6	0	0.0	353
Imports	*18*	*1.2*	*1505*	*4.9*	*6*	*0.3*	*84*
Total	**1463**		**30446**		**2250**		**21**

TABLE 5.10 SUMMARY OF THE HOLLOW WAY FILL (1005) ASSEMBLAGE BY FABRIC (EXCLUDING RESIDUES)

4th to 5th century. Shell-tempered ware has been recorded on a variety of late Roman rural sites in Worcestershire and further afield, but it has only rarely been found on sites in Worcester itself (Griffin 2005a; Griffin 2005b; Timby 2004; Rees forthcoming), It was found nearby at 1 The Butts, in a late Roman well (Evans, Chapter 3), but was not present in the 'dark earth' assemblage from 8–12 the Butts (Evans, Chapter 4). The well contained nine 4th-century coins, all from the upper fills, the latest dating to 364–78 (Napthan, Chapter 4). The rarity of wheelmade, shell-tempered ware in Worcester has often provided supporting evidence for the idea that sites were abandoned sometime between the late 3rd or early 4th century, as at Sidbury (Darlington and Evans 1992, 32), Deansway (Bryant and Evans 2004, 280), Farrier Street (Dalwood *et al* 1994, 97), and the New Police Station and Magistrates's Court sites in Castle Street (Edwards *et al* 2002, 129). At Sidbury, the latest deposits are dated by coins to *c* 270–73

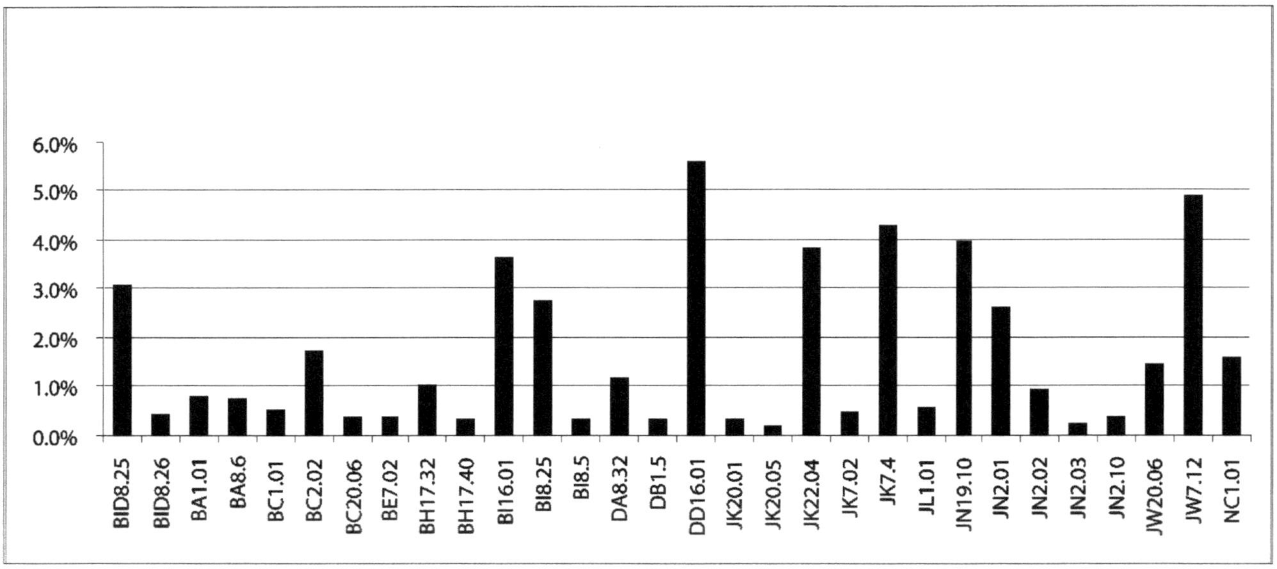

Concordance of Form codes and figure numbers: B/D8.25 , Fig. 5.21, 57; B/D8.26 , too fragmentary for illustration, cf BA8.6? ; BA1.01 , Fig. 5.21, 67 ; BA8.6 , Fig. 5.21, 73 ; BC1.01 , Fig. 5.19, 24, Fig. 5.20, 33, Fig. 5.21, 66; BC2.02, Fig. 5.21, 65; BC20.06, Fig. 5.19, 23 ; BE7.02, Fig.5.21, 68; BH17.32, Fig. 5.19, 18; BH17.40, Fig. 5.19, 21 ; BI16.01 , Fig. 5.21, 70 -1; BI8.25, Fig. 5.20, 39; BI8.5 , Fig. 5.21, 72 ; DA8.32, Fig. 5.20, 40; **DB1.5**, Fig. 5.20, 41 ; DD16.01 , Fig 5.20, 58; JK20.01 , Fig. 5.21, 61 ; JK20.05, Fig. 5.21, 60 ; JK22.04, Fig. 5.20, 46 -7; JK7.02, Fig. 5.21, 59 ; JK7.4, Fig. 5.21, 52 -4; JL1.01, Fig. 5.20, 44 ; JN19.10, Fig. 5.19, 4, Fig. 5.20, 45; JN2.01 , Fig. 5.19, 6; JN2.02, Fig. 5.19, 8 ; JN2.03, Fig. 5.19, 7 ; JN2.10, Fig. 5.19, 9 ; JW20.06 , Fig. 5.19, 11; JW7.12 , Fig. 5.19, 14 -15; NC1.01 , Fig. 5.19, 3;

FIGURE 5.21 DIAGNOSTICALLY LATE VESSEL FORMS (% RIM EVE) FROM 1005

(Darlington and Evans 1992, 32) and at Deansway, where coin evidence indicated activity continuing to at least AD 364–78 (Bryant and Evans 2004, 280), only three out of 140,000 sherds were in this fabric. Small quantities have been noted in dark earth deposits elsewhere in Worcester, and in the late Roman assemblage from the County Education Offices (Dalwood *et al* 1997, fig. 6).

The evidence from other towns in the region, for example Alcester (Evans 1994, 146), supports a late 4th-century date for its use. It was well represented in the late assemblage from Evesham Road, Pershore, for which a *terminus post quem* of *c* 390 has been suggested (Griffin 2005a, 11–12, fig. 17). At Gloucester it is associated with a coin hoard dated to after 388 (Hassall and Rhodes 1975, 85–86). It is also a diagnostic characteristic of Phase 4 (post AD 355) deposits at Bays Meadow villa in Droitwich (Barfield 2006). Further north, at Wroxeter, shell-tempered wares first appear in late 4th-century contexts, becoming increasingly common in later phases (Symonds 1997, figs 366 and 372). They are also characteristic of late deposits at *Segontium* in north Wales (Webster 1993).

The coins associated with the hollow way fill (1005) are a little earlier than the above examples: two dating to AD 330–35 and one to AD 337–40 (Nogueira, this chapter). However, a coin from the evaluation Trench 1 (F108E) has indicated activity on the site to at least AD 346–50 (Nogueira, this chapter). While this could indicate a slightly earlier date for the use of shell-tempered ware in Worcester, it is perhaps more likely that the issue date of the coins does not provide a secure date for the use and deposition of the pottery. There is a similar mismatch of evidence at the Evesham Road, Pershore site, where the ceramic *terminus post quem* of *c* AD 390 (Griffin 2005a, 12) is not supported by the issue dates, AD 354–64, of the latest coins (*ibid*, table 6). Jars were the only form recovered in this ware (Fig. 5.20, 59–62), with parallels indicating a mid to late 4th-century date (Fig. 5.20, 59–60).

Oxfordshire wares

Oxfordshire wares (WHEAS fabric 29) provided further reliable dating evidence (Fig. 5.20; 64–73, 77–78). Red colour-coated wares included a beaker and bowls dating securely to the 4th century (Fig. 5.20, 64–65, 67–69) as well as forms dating more broadly to 240–400+ (Fig. 5.20, 66–72). The latest form (Fig. 5.20, 69) dated to the second half of the 4th century (340–400+). A carinated body sherd with linear stamped decoration (not illustrated) also indicated a date in at least the second quarter of the 4th century (Young 2000, fig. 61, C70 or C71).

The Oxfordshire parchment wares (WHEAS fabric 40) included a painted bowl (Fig. 5.20, 73) of a type produced from *c* 240, but particularly popular in the 4th century. This form was also noted at 1 The Butts (Evans, Chapter 3). Oxfordshire mortaria included white ware (WHEAS fabric 33.1), white slipped (WHEAS fabric 33.2) and red colour-coated ware (WHEAS fabric 33.3). The mortaria included the principal Oxfordshire mortarium produced from *c* 300 (Fig. 5.20, 76), as well as a longer lived type (Fig. 5.20, 78) and more typically 3rd-century types (Fig. 5.20, 75, 77). Mortaria from Mancetter-Hartshill were also present (WHEAS fabric 32). Though few rims were included. Small quantities of Nene Valley colour-coated ware were present (WHEAS fabric 28). Forms included: a 'Castor Box' (Fig. 5.20, 63), a rouletted body sherd, probably from a jar or bottle (cf. Howe *et al* 1980, fig. 6, 69) and the base of a flagon, all dating to the 4th century, and a body sherd decorated with barbotine over the slip, dating to at least the late 3rd century (Howe *et al* 1980, 8). A single sherd in a micaceous, colour-coated ware is thought to be Southwestern brown slipped ware (WHEAS fabric 151) from Gloucestershire (Lee *et al* 1994, 12). This late 4th-century ware has been noted at Alcester and other sites in the region (Evans 1994, 147–48, fig. 67), mainly associated with beakers. This ware is very similar in style, decoration and fabric to Oxfordshire ware, making these fabrics difficult to differentiate (Lee *et al* 1994, 12). It is possible, therefore, that some additional sherds might have been recorded as Oxfordshire ware.

Dorset Black Burnished ware (BB1)

Dorset Black Burnished ware (BB1: WHEAS fabric 22) was well represented. Most of the forms were broadly late 3rd- to 4th-century types: splayed rim jars decorated with obtuse cross hatch (Fig. 5.20, 52–54) or less typical burnished motifs (Fig. 5.20, 55–56), oval fish dishes (Fig. 5.20, 58) and flanged rimmed bowls or dishes (Fig. 5.20, 57). The evidence from Alcester suggests that the latter ceased to be used in the region by the end of the 4th century (Evans 1994, 146). None were present at 1 The Butts (Evans, Chapter 3). One body sherd, from a jar, was perforated for repair. At Wroxeter, signs of repair are first noted in assemblages dated *c* 367–450 (Barker *et al* 1997, 240). This sherd, however, was decorated with square lattice which dates the vessel, though not necessarily the repair, to *c* 160–280.

Pink grogged ware

Further evidence of late 3rd- to 4th-century activity was provided by large storage jars in pink grogged ware (WHEAS fabric 17; Booth and Green 1989). The forms (Fig. 5.19, 44–45) are typical of the late products of this industry, which appears to have been based in the Milton Keynes area (*ibid*, fig. 3). This ware has been found in late contexts elsewhere in the county, for example at Childswickham (Timby 2004).

Severn Valley ware

Many of the Severn Valley ware forms (WHEAS fabric 12, 12.1) supported the evidence from these well-dated, traded wares. A number were diagnostically 4th-century types, with parallels at the late kiln site in Malvern (Peacock 1965–67) or Bays Meadow villa in Droitwich (Barfield 2006). These included tankards (Fig. 5.18, 3), jars and/ or bowls (Fig. 5.18, 11, 13–14). Other jars and bowls dated broadly to the late 3rd to 4th century (Fig. 5.18, 6–9, 15, 18–21, 23; Fig. 5.19, 34). Various longer lived forms can, by association, be assumed to be late examples of their type (Fig. 5.18, 4, 10, 17, 22, 24; Fig. 5.19 33).

Other wares

The other coarse wares (WHEAS fabrics 14, 15, 21.3, 149 and 19) included a number of copies of late 3rd- to 4th-century BB1 types (Fig. 5.19, 37, 39, 43, 46, 47), as well as longer lived BB1 forms (Fig. 5.19, 38, 42). Dishes in the coarse sandy ware (WHEAS fabric 15) had parallels at the 4th-century Hygienic Laundry production site in Malvern (Fig. 5.19, 40–41). The reduced fabric with blackened surfaces (WHEAS fabric 149) is paralleled in late Roman assemblages from Evesham Road, Pershore (Griffin 2005a, 13, Fabric 149) and Wyre Piddle (Laura Griffin pers. comm.), and was recorded at 1 The Butts (Evans, Chapter 3). The fine, micaceous grey ware (WCM 21.3), was also noted at 1 The Butts, where forms included an everted rim jar paralleled in a post-Roman context at Droitwich (Rees 1992, fig. 31, 18).

The pottery catalogue from the hollow way fill (1005)

The following catalogue relates to material recovered from the hollow way fill (1005) which must be considered a post-medieval context (Phase 3) which contains a significant quantity of 4th-century residual Roman material. The catalogue entries are numbered 1–78, with forms 1–31 illustrated in Fig. 5.18, forms 32–51 illustrated in Fig. 5.19, and forms 52–78 illustrated in Fig. 5.20. The entries are listed under their Worcestershire County Fabric Series with individual fabric, form, description and context.

WHEAS 3, MAL RE A (N02.1): Handmade Malvernian ware

1 L. Lid handle, not joining the rim below, but possibly from the same vessel.

2 LAA16.00. Concave lid with a plain rim. Diam. 180mm, (9%).

WHEAS 12, SVW OX 2 (O02.1): Severn Valley ware, plain

3 NC1.01. Fragmentary, splayed rim from a tankard, a typical 4th-century type (Webster 1976, fig. 7, E44). Diam. 100mm (10%).

4 JN19.10. Triangular rim from a narrow-mouthed jar, with a fairly short neck. Similar rims are dated by Webster broadly to 2nd to 4th century (1976, fig. 1, A4). The short neck on this example might indicate a date towards the end of this range. Diam. 120mm, (28%).

5 JN20.01. Rounded, slightly overhanging rim from a narrow-mouthed jar. Diam. 130mm (26%).

6 JN2.01. 'Bifurcated' or pulley rim from a narrow-mouthed jar. The form is dated by Webster (1976, fig. 3, A9) to the 3rd to 4th century. Similar rims are illustrated from the late 4th-century kiln at Meole Brace, Shropshire (Evans *et al* 1999, fig. 8, 7), and from Droitwich, Bays Meadow Phase 3, dated to *c* AD 289–96 (Barfield 2006, fig. 97, 118). Diam. 90mm, (16%).

7 JN2.03. Narrow-mouthed jar with a variant on the pulley rim, the upper bead being more pronounced; similar to an example from the late 4th-century kiln at Meole Brace, Shropshire (Evans *et al* 1999, fig. 8, 9). Diam. 130mm, (5%).

8 JN2.02. Narrow-mouthed jar with a variant on the pulley rim, the lower bead being more pronounced. Diam. 90mm (21%).

9 JN2.10. Narrow-mouthed jar with a triple beaded pulley rim. Diam. 140mm, (8%).

10 JW19.10. Triangular rim from a wide-mouthed jar. Similar rims are found on 2nd- to 3rd-century jars published by Webster (1976, fig. 4, C22), though the short neck on this example supports a 4th-century date (cf. *ibid*, fig. 7, C31–33). Diam. 220mm, (14%).

11 JW20.06. Hooked rim from a wide-mouthed jar, with a short neck. The form is dated by Webster to the 4th century (*ibid*, fig. 7, C31–33). It is associated with the 4th-century Hygienic Laundry kiln in Malvern (Peacock 1965–67, fig. 3, 52–54, fig. 4, 55–60), and the late 4th-century kiln at Meole Brace, Shropshire (Evans *et al* 1999, fig. 8, 11–13). It is also known from the Malvern, Newland Hopfield kiln site (Evans *et al* 2000, fig. 24, JWM17, 18), and from Worcester, Deansway (Bryant and Evans 2004, fig. 154, 5). Diam. 280mm, (5%).

12 JW7.03. Cavetto rim from a small jar. Diam. 100mm, (18%).

13 JW7.03. Rim from a short-necked jar or bowl of near equal to narrow girth. No good parallels are published by Webster. They are dated by Rawes to the 2nd century (Rawes 1982, fig. 4, 68–69, 45), but 4th-century parallels can also be found in the Malvern, Hygienic Laundry kiln assemblage (Peacock 1965–67, fig. 3, 35). Diam. 230mm, (16%).

14 JW7.12. Out-curving rim from a neckless jar or bowl. A similar form is illustrated from Droitwich Bays Meadow, Phase 4, post-dating *c* AD 355 (Barfield 2006, fig. 100, 204). Diam. 210mm, (19%).

15 JW7.12. Out-curving rim from a neckless jar or bowl with one or possibly two handles. A similar handled bowl is illustrated from Droitwich Bays Meadow, Phase 3, dated to *c* 289–96 (*ibid*, fig. 97, 115). Diam. 190mm, (12%).

16 JW19.11. Fragmentary, squared off rim from a wide-mouthed jar. Diam. 290mm, (12%).

17 BC21.03. Rim from a medium- to large-sized flanged bowl with an internal lip and curving sides (Webster 1976, type F, fig. 8, 45–46, 50). The form appears to have been produced throughout the Roman period (*ibid*, 31–33). It is known from kiln sites in the Malvern area (Evans *et al* 2000, fig. 26, BT7, BT8; Tomber 1980, fig. 16, 246, fig. 17, 248), and from other sites in Worcester (Bryant and Evans 2004, fig. 157, 7; Darlington and Evans 1992, fig. 23, 10), though predominantly from 2nd- and 3rd-century contexts. However, a parallel is illustrated from the Droitwich, Bays Meadow Phase 4 assemblage, post-dating *c* AD 355 (Barfield 2006, fig.100, 199). Diam. 230mm, (15%).

18 BH17.32. Flanged, reed rim from a medium- to large-sized bowl, with a raised central bead and an internal lid seat and lip. The type is dated by Webster to the late 3rd to 4th century (1976, fig. 9, F53), based on the presence of similar forms at Bays Meadow villa, Droitwich (Barfield 2006, fig. 102, 317) and the Grit Farm kiln site, Malvern (Webster 1976, 33: Malvern kiln III; Tomber 1980, 37–38). It is also known from the Malvern, Newland Hopfields kiln site (Evans *et al* 2000, fig. 28, BT31). Diam. 190mm (14%).

19 BH17.30. Flanged, reed rim from a medium to large sized bowl with an internal lip and a raised central bead. The form is known at Deansway, Worcester (Bryant and Evans 2004, fig. 155, 13), and a similar form is published from Droitwich Bays Meadow phase 3, dated to *c* 289–96 (Barfield 2006, fig. 97, 112). Diam. 230mm (10%).

20 BH17.30. Variant of above with broader central bead. Diam. 240mm (10%).

21 BH17.40. Flanged, medium- to large-sized bowl with a grooved rim and an internal lip. The form is dated by Webster to the 3rd century (1976, fig. 9, G57), though a parallel is illustrated in the Phase 4 assemblage from Droitwich Bays Meadow, post-dating *c* 355 (Barfield 2006, fig. 100, 201). Diam. 230mm (7%).

22 BH17.20. Flanged, medium- to large-sized bowl, with a high flange and a pronounced internal lip. Diam. 220mm (5%).

23 BC20.06. Small bowl with a hooked rim, similar to a vessel illustrated from phase 3c, *c* AD 289–96, at Droitwich Bays Meadow (Barfield 2006, fig. 98, 141). Diam. 140mm, (8%).

24 BC1.01. Small bowl with an upright, beaded rim, Webster (1976, fig. 9, 61) type I, reminiscent of samian form 37, produced from the mid to late 1st century onwards. A similar form was produced at the Malvern, Newland Hopfields kiln site (Evans *et al* 2000, fig. 30, Type 5, B64). Diam. 130mm (5%).

25 BA. Body sherd from a carinated bowl, decorated with white painted lines.

26 BA? Body sherd, possibly from a carinated bowl, decorated with white painted lines and dots.

27 LAA19.10. Thickened rim from a concave lid, not closely datable. Diam. 240mm (13%).

28 LAA16.00. Plain rim from a concave lid, not closely datable. Diam. 150mm (6%).

29 LAB19.10. Slightly beaded rim from a convex lid, not closely datable. Diam. 18mm (3%).

30 LAB16.00. Rim from a convex lid, not closely datable. Diam. 140mm (30%).

31 L. Handle from a lid, crudely finished with string marks visible.

WHEAS12.1, SWV R (G04): Reduced Severn Valley ware

32 JW7.12. Out-curving rim from a neckless jar or bowl, cf. Fig. 5.18, 14 above. Diam. 240mm (3%).

33 BC1.01. Bead rim bowl, Webster type I (Webster 1976, fig.9, 61, 64), similar to types produced at Newland Hopfields, Malvern (Evans *et al* 2000, fig. 30, Type 5, BT64). Not a closely dated form. The illustrated example is decorated with cross hatch and has a deliberately blackened surface on the pale grey fabric, perhaps copying BB1. Diam. 180mm (3%).

34 BH17.30. Flanged, reed rim from a medium sized large bowl with an internal lip, and a raised central bead. See Fig. 5.18, 19 above. Diam. 240mm (14%).

WHEAS 12.2, SVW ORG OX (O03.1): Organic-tempered Severn Valley ware, oxidised

35 JN7.01. Splayed, bead rim, similar to Webster A7, dated broadly to the 2nd to 3rd century (Webster 1976, fig. 2). Possibly a waster. Residual? Diam. 160mm (6%).

WHEAS 13, (O06.18): Sandy oxidised ware

36 F7.01. Fragmentary rim from a very slightly cup mouthed flagon. Diam. 60mm (17%).

WHEAS 14 (G012.3): Reduced sandy ware, fine

37 JK7.40. Fragmentary splayed rim, probably from a BB1 copy jar. Diam. 130mm (9%).

38 DB16.01. Copy of a BB1 plain-rimmed dish, type 20 (Seager Smith and Davies 1993, fig. 123), not a closely datable form. Diam. 160mm (8%).

WHEAS 149 (G012.2): Worcestershire imitation Black Burnished ware

39 BI8.25. Flanged bowl, copy of a late 3rd- to 4th-century BB1 type (*ibid*, fig. 124, type 25). Diam. 150mm (34%).

WHEAS 15 (G012.1): Reduced sandy ware, coarse

40 DA8.32. Flanged dish with crude pattern burnish internally. Copy of a BB1 type. Similar forms were noted at the 4th-century Hygienic Laundry production site in Malvern, though it was not certain that they were kiln products (Peacock 1965–67, fig. 4, 72). These were also is a sandy fabric. Parallels in fabric WHEAS 15 can be found in the phase 3c, *c* 289–96, assemblage from Droitwich Bays Meadow (Barfield 2006, fig. 99, 178–80) and from Worcester, Deansway (Bryant and Evans 2004, fig. 163, 13). Similar, crude decoration can be seen on other reduced Severn Valley ware and Malvernian dishes found in Worcester (*ibid*, fig. 163, 13; fig. 164, 7). Diam. 230mm (13%).

41 DB1.5. Dish with pinched finger tip decoration on the rim. Similar forms were noted at the Malvern, Hygienic Laundry site (Peacock 1965–67, fig. 4, 70). An example is illustrated from the Droitwich, Bays Meadow, Phase 3 assemblage, dated to *c* AD 289–96 (Barfield 2006, fig. 97, 126). Diam. 200mm (7%).

42 DB16.01. Copy of a BB1 plain-rimmed dish (Seager Smith and Davies 1993, fig. 123, type 20), not a closely datable form. Abraded, but appears to have had a deliberately blackened surface. Diam. 160mm (10%).

43 B/ D8.26. Fragmentary rim of a BB1 copy, dropped flange bowl. Diam. 200mm (9%).

WHEAS 17, PNK GT (F018): Pink, grog-tempered ware

44 JL1.01. Abraded rim from a large, coil-built storage jar. Though the precise form of the rim is uncertain, the jar type is a typical of the late 3rd- and 4th-century forms produced in this fabric (Booth and Green 1989, fig. 1, 1 or 2, 82). A similar form was noted at Droitwich Bays Meadow in phase 3, dated to *c* AD 289–96 (Barfield 2006, fig. 97, 128). Diam. 300mm, (12%).

45 JN19.10. Necked, narrow-mouthed jar with a triangular rim, decorated with a notched cordon at the base of the neck. Probably dating to the late 3rd to 4th century. A similar form in this fabric is published from Alcester (Booth and Green 1989, fig. 2, 12). Diam. 190mm (15%).

WHEAS 19, (N04) Wheelmade Malvernian

46 JK22.04. Fragmentary, splayed rim from a jar. Similar to type recorded at the Hygienic Laundry kiln site in Malvern, dated to the 4th century (Peacock 1965–67, fig. 3, 20–21, 28). Diam. 180mm (5%).

47 JK22.04. Fragmentary, splayed rim from a jar, smaller variant of above. Diam. 140mm (20%).

48 BC21.02. Curving sided bowl or dish with a T-shaped, flange rim. A common Severn Valley ware type. Generally dated to the 2nd to 3rd century (Evans *et al* 2000, fig. 26, Type 2), but known from a 3rd- to 4th-century context at Astley (Walker

1959, fig. 13, 101), and similar to a Phase 4 (post *c* 355) form from Droitwich Bays Meadow (Barfield 2006, fig. 100, 199). Diam. 200mm (16%).

49 BCB1.01. Shallow bowl or dish with curving walls and a bead rim. Webster type K74, not closely dated (Webster 1976, fig. 10, 36). A similar form was produced at Newland Hopfields (Evans 2000, fig. 30, type 6.2, BT69). Diam. 140mm (8%).

50 BB8.31. Large, straight sided bowl with a T-shaped, flange rim, variant of a Severn Valley ware type (see Fig. 5.18, 17 above). Broadly dating from the 2nd century onwards. Diam. 350mm (16%).

51 LAA19.11. Fragmentary, hooked rim from a lid. Diam. uncertain (3%).

WHEAS 22, DOR BB1 (B02): Dorset Black Burnished ware

52 JK 7.4. Rim from a jar with a markedly splayed rim, decorated with a band of obtuse lattice cross hatch, the upper edge of which is delineated by a burnished groove. A typical late 3rd- or 4th-century type (Seager Smith and Davies 1993, 231, fig. 122, type 3). Diam. 170mm (9%).

53 JK7.4. Variant of above. Diam. 180mm (3%).

54 JK7.4. Variant of above. Sooted externally and inside rim. Diam. 170mm (8%).

55 JK. Decorated body sherd from a jar, similar to decoration type D8 (Seager Smith and Davies 1993, fig. 132).

56 JK. Decorated body sherd from a jar, a variant of obtuse cross hatch burnish.

57 B/ D 8.25. Fragmentary rims from a late 3rd- to 4th-century, flange-rimmed bowl or dish (Seager Smith and Davies 1993, fig. 124, type 25). Decorated externally with burnished, interlocking arcs. Diam. 160mm (4%).

58 DD16.01. Part of an oval 'fish' dish. An uncommon BB1 form, found outside Dorset in late 3rd- to late 4th-century contexts (Seager Smith and Davies 1993, fig. 123, type 21, 233; Gillam 1976, 78, fig. 6, 85). With graffiti (see Tomlin, this chapter). Diam. NA.

WHEAS 23, HAR SH (H04): Shell-tempered ware

59 JK7.02. Outcurving, slightly beaded rim from a jar. A similar form, dating to the mid to late 4th century, is published from Alcester (Lee *et al* 1994, fig. 2, R42). The illustrated sherd has sooting externally. Diam.180mm (10%).

60 JK20.05. Slightly overhanging rim from a jar. A similar form is published from Alcester, from a context dating to after *c* 353 (Lee *et al* 1994, fig. 2, R43, phase XIII). Sooted externally. Diam. 160mm (4%).

61 JK20.01. Hooked rim from a jar. Sooting externally and inside rim. Diam 210mm (7%).

62 J. Decorated sherd from a shelly ware jar.

WHEAS 28.1, LNV CC (C03): Lower Nene Valley colour-coated ware

63 LC1.04. Rim from a 'Castor Box' lid dating to the 4th century (Howe *et al* 1980, fig. 7, 89). Diam. 170mm (8%).

WHEAS 29, OXF RS (C07): Oxfordshire red colour-coated ware

64 BK. Body sherd from a beaker with a near vertical neck and a globular body, with white painted decoration. Possibly a Young type C27, usually 4th century (Young 2000, fig. 55, 154).

65 BC2.02. Double bead rim from a full bellied bowl. Young (2000 fig. 60, 162) type C71, dated to the 4th century, 300–400+. Related forms (Young 2000, fig. 61, C73) are included in the post *c* 355 assemblage from Droitwich Bays Meadow (French 2006). Diam. 160mm (38%).

66 BC1.01. Hemispherical bowl with a bead rim. Plain apart from an incised groove below the girth. Young (2000, fig. 60, 160) type C55 dated 240–400+. Diam. 170mm (6%).

67 BA1.01. Wall sided, bead rim, carinated bowl. Young type (2000, fig. 64, 166) C81, dated consistently to the 4th century, 300–400. Diam. 120mm (18%).

68 BE7.02. Fragmentary, out-turned rim from a necked bowl with a full, curved body, Young (2000, fig. 62, 164) type C75.1. One of the most common Oxfordshire colour-coated types, produced from the second quarter of the 4th century, 325–400+. Diam. 170mm (6%).

69 BE7.02. Fragmentary rim from a necked bowl, with a beaded, everted rim; Young (2000, figs 62–63, 166) type C78, dated to 340–400+. Diam. Uncertain (2%).

70 BI16.01. Flanged bowl, copying Dr38. Young (2000, fig. 59, 160) type C51, the most common Oxfordshire colour-coat type, made throughout the production period, 240–400+. Diam. 140mm (22%).

71 BI16.01. Stubby rim from a flanged bowl. Variant of Young (2000, fig. 59, 160) type C51, 240–400+. Diam. 110mm (32%).

72 BI8.5. Shallow bowl with an out-turned rim, up-turned at the tip. Young (2000, fig. 58, 158) type C49 dated 240–400+. Diam. 220mm (7%).

WHEAS 40, OXF PA (P08): Oxfordshire parchment ware

73 BA8.6. Wall sided bowl, moulded at the rim and carination, and decorated with red, horizontal bands. Young (2000, fig. 27, 87) type P24, the most common Oxfordshire parchment ware form. The form was produced from 240–400+, but was particularly popular in the 4th century. An example was noted in the Phase 4 (post *c* 355) assemblage at Droitwich Bays Meadow (French 2006). Diam. 230mm (11%).

WHEAS 32, MAH WH (M02): Mancetter-Hartshill mortaria

74 M4.0. Hammer head rim, with red painted decoration on the flange. 3rd century. Diam. 260m (3%).

WHEAS 33.1, OXF WH (M04c): Oxfordshire white mortaria

75 ME8.015. Fragmentary rim from an Oxfordshire type M17 mortarium (Young 2000, fig. 21, 72), with an upstanding rim, and a flange hooked under at the tip. The form is dated by Young (2000) to 240–300, and is thought to be residual in 4th-century contexts. Diam 300mm (5%).

76 ME8.15. Fragmentary rim from an Oxfordshire type M22 mortarium (Young 2000, fig. 23, 76) with a squat flange, grooved at tip. The form dates broadly to 240–400+, and was the principal Oxfordshire mortarium produced from *c* 300. This was the most common Oxfordshire mortarium type at Droitwich Bays Meadow, particularly in the Phase 4 assemblage (Barfield and Hartley 2006). Diam. 290mm (6%).

WHEAS 33.2, OXF WS (M04b): Oxfordshire white slipped mortaria

77 ME8.015. Rim from a Young type WC 4 mortarium (Young 2000, fig. 38, 120). The form copies type M17 and is assumed to have a similar date range of AD 240–300. Diam. 210mm (12%).

WHEAS 33.3, OXF RS (M04a); Oxfordshire red colour-coated mortaria

78 MB4.10. Rim from a red colour-coated Oxfordshire mortarium (*ibid*, fig. 67, type C97, p. 173). A common and long lived Oxfordshire form, dating to AD 240–400. Diam. 180mm (18%).

Functional composition

Condition and character

The pottery from hollow way layer 1005 was relatively well preserved. Sooting and limescale survived on a few sherds (24 sherds and 20 sherds respectively), but most sherds were abraded to some degree. The assemblage included numerous sizeable fragments, with a fairly high average sherd weight (20g, excluding the four residual sherds of amphora which had an average sherd weight of 353g). This is the same as the average weight of sherds sealed within the well at 1 The Butts (20g; Evans, Chapter 3), but higher than the average weight of sherds in the 'dark earth' deposit at 8–12 The Butts (12g; Evans, Chapter 4).

It is suggested that the general condition and character of the assemblage points to a pattern of disuse and dumping which is consistent with the evidence from 1 The Butts (Napthan, Chapter 3) and that the assemblage derives from a dump or midden rather than a more defined activity area. This makes interpretation of the pottery use more complex. In the discussion below, the assemblage is compared with other late Roman assemblages from nearby excavations, particularly from 1 The Butts and 8–12 The Butts (Chapters 3 and 4), in order to assess how representative the group might be of late Roman pottery use in Worcester.

Vessel classes and vessel function (Fig. 5.22)

The range of vessel classes from the hollow way fill (1005) is quantified in Fig. 5.22a, while Fig. 5.22b and c show the range of vessel classes from two nearby late Roman groups: the group from 1 The Butts (Chapter 3) derives from the fill of a well; the group from 8–12 The Butts (Chapter 4) derives from a 'dark earth' deposit. While the group from the fill of the well could be a coherent functional group, potentially derived from a dwelling associated with the well, the group from 8–12 The Butts and from layer 1005 at 14–24 The Butts are more likely to have accrued in a midden, or to have lain on the surface over a long period of time, and are perhaps therefore more mixed functional and chronological groups.

Fig. 5.22a shows that jars and bowls were the two most commonly represented classes of pottery vessel in layer 1005 at 14–24 The Butts. Comparison with the groups from the well at 1 The Butts (Fig. 5.22b) and the 'dark earth' at 8–12 The Butts (Fig. 5.22c) indicates that jars and bowls are well represented in all three groups, although the indeterminate class of bowls/ dishes complicates more exact comparison.

Jars were certainly the most common class in all the late assemblages. As shown on Fig. 5.22a–c, the jars were subdivided into five groups for quantification. Severn Valley ware wide-mouthed storage jars were the largest group in layer 1005 and in the 'dark earth' deposit at 8–12 The Butts. They were, in contrast, very poorly represented in the well at 1 The Butts. A range of variations in these jars was noted in layer 1005 (Fig. 5.18, 10–16 and Fig. 5.19, 32), the most common types being nos 13 (37.5% of this class) and 14 (20%). Narrow-mouthed storage jars were also fairly common. The majority (97%) was in Severn Valley ware (Fig. 5.18, 4–9 and Fig. 5.19, 35), with a single vessel in pink grog-tempered ware (Fig. 5.19, 45). Large storage jars were the least common type. These were all in fabrics with coarser tempers, necessary when firing thick

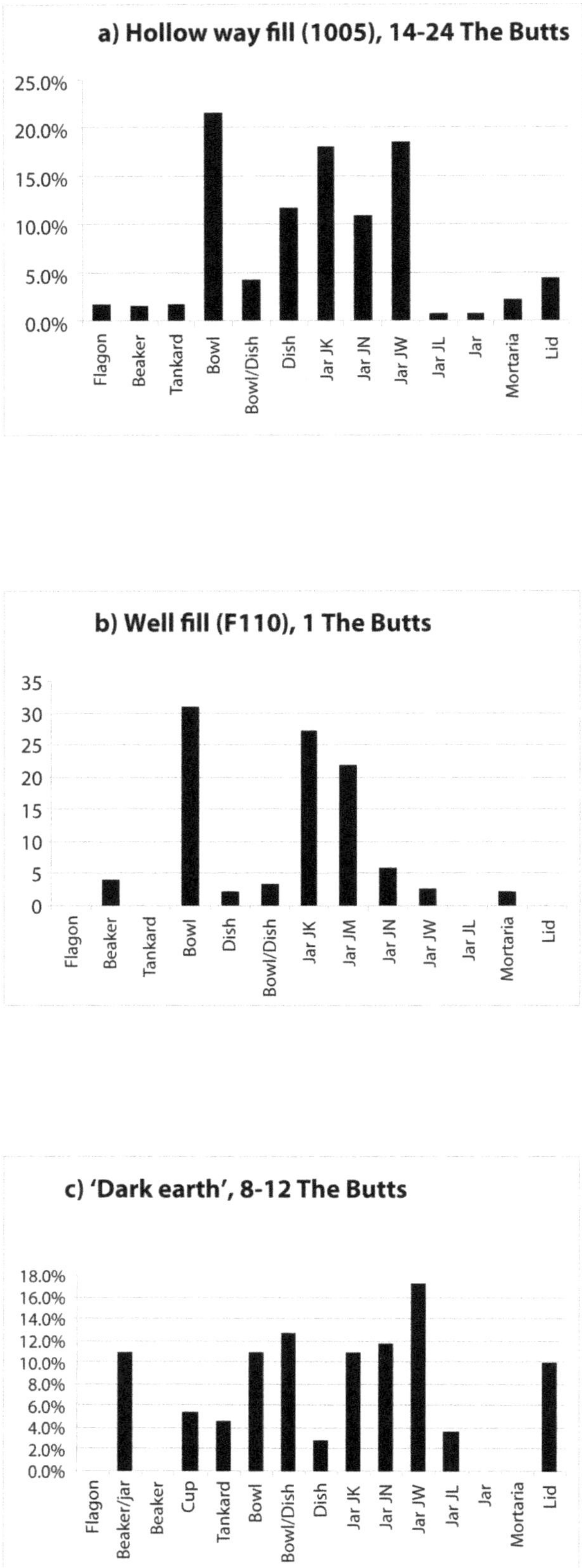

FIGURE 5.22 LATE ROMAN VESSEL CLASSES (% RIM EVE)

walled vessels. A range of sources was represented, mainly fairly local (WHEAS 12.2, 64%; WHEAS, 3 4%) but some indicating longer distance trade of the commodities they contained (WHEAS 17, 32%).

Coarsely-tempered jars classified as 'cook pots' were common in layer 1005, and particularly so in the well group. Most of those from 1005 (70.5%) were in BB1. Of these 28% had characteristically late, splayed rims (Fig. 5.20, 52–54; Seager Smith and Davies 1993, fig. 122, type 3), 23.5% moderately splayed rims (*ibid*, type 2), and 3% upright rims (*ibid*, type 1). Both of the latter may include residual vessels. Unfortunately, the majority (45%) were too fragmentary to be attributed to precise form types. Copies of late BB1 cook pots occurred in wheelmade Malvernian ware (21%; Fig. 5.19, 46–47) and local sandy wares (3.5%; Fig. 5.19, 37). Shell-tempered jars are also included in this category (5%; Fig. 5.20, 59–62), on the basis that six sherds in this fabric had sooting. This contrasts with the evidence from the well group, amongst which only the BB1 and sandy ware jars had sooting (Evans, Chapter 3). Evidence for use in cooking was also noted on BB1 jars from this site: nine sherds had sooting and 19 limescale.

Lids, presumably used in conjunction with the jars, were present both in 1005 and the 'dark earth' assemblages at 8–12 The Butts, but not in the well assemblage at 1 The Butts (Fig. 5.22a–c). Most (72%) were in oxidised Severn Valley ware, presumably used in conjunction with the various storage jars. Others (17%) were in Malvernian fabrics, and would have been suitable for cooking or storage vessels. The remainder was from Nene Valley colour-coated 'Castor boxes' (Fig. 5.20, 63).

Bowls were the second most common class in layer 1005 and in the well group at 1 The Butts (Fig. 5.22a and b). The lower proportion in the 'dark earth' at 8–12 The Butts might reflect that more sherds here were only broadly classifiable as bowl/ dish. Oxfordshire red colour-coated table wares made up 78% of the bowls in the well, though bases from BB1 bowls/ dishes were also noted. In layer 1005, Severn Valley ware bowls were most common (36% of the bowls by rim EVE). Here 21 % were utilitarian, medium to large bowls with flange rims (Fig. 5.18, 17–22), presumably used in food preparation, while 4 % were smaller bowls, possibly used as table wares (Fig. 5.18, 23–24). The remainder were less diagnostic forms.

Figure 5.23 plots the rim diameters for all bowls. Three size ranges are evident: small; small to medium; and medium to large. The smaller group centres on 110mm, the small–medium group on 140–60mm, and the larger group on 230mm. The Oxford ware bowls, representing 31% by rim EVE, were mainly in the smaller group, with diameters ranging from 90–220mm. The emphasis on Oxfordshire bowls in the well could thus explain the bias towards smaller bowls in that group (Fig. 5.23b). One base sherd from a bowl or dish had graffiti, indicating that it had been the 'property of Rufus' (Tomlin, this chapter). Nene

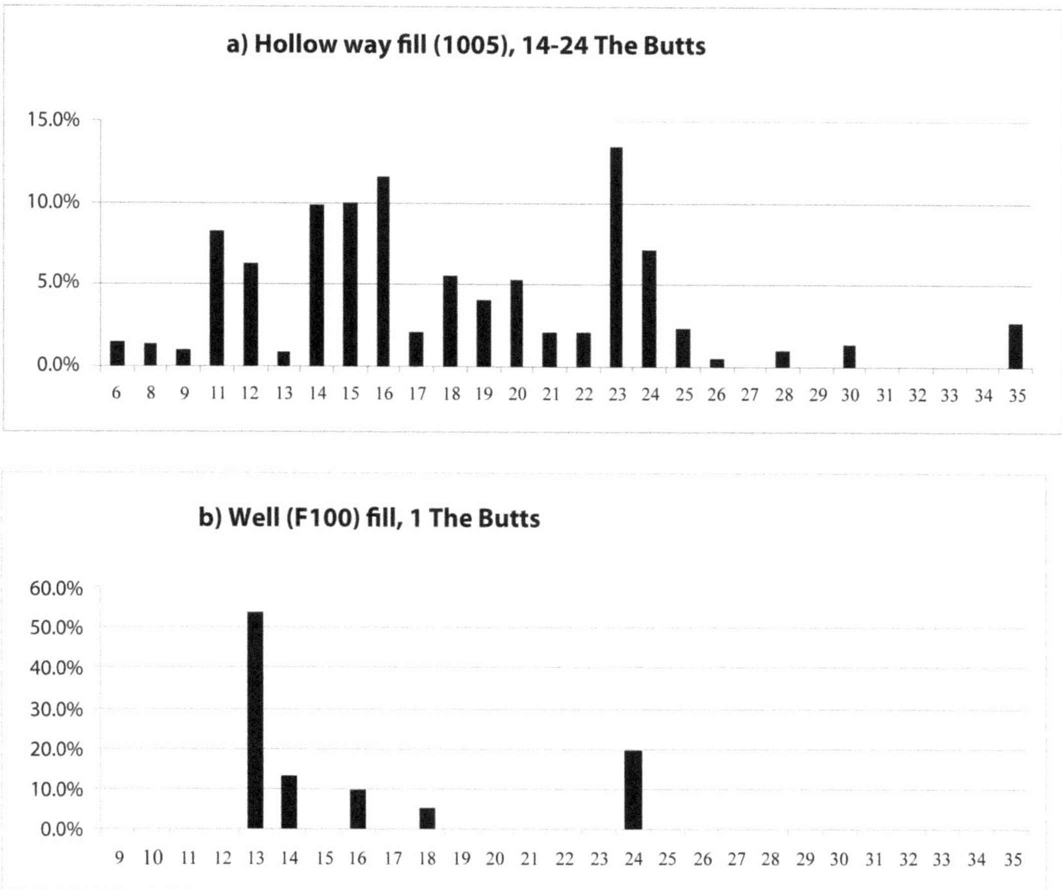

FIGURE 5.23 RIM DIAMETERS FOR BOWLS (% RIM EVE)

Valley colour-coated wares accounted for a further 3%, again occurring in smaller sizes. The remainder was bowls or bowls/ dishes in BB1 (16.5%), or copies of BB1 forms in wheelmade Malvernian ware (6%) or other local fabrics (7%). No shell-tempered bowls were recovered on this or the other Butts sites, though they have been recorded in late assemblages elsewhere in Worcestershire (Griffin 2005a, fig. 17, 25–28).

Dishes appear to be better represented in the layer 1005 assemblage than in the other two groups (Fig. 5.22a–c). The majority were in BB1 (76% of the dishes), including the typically late 'fish' dish variant (Fig. 5.20, 58). One of the BB1 dishes bore another graffito, crudely inscribed underneath the base, and also thought to be a mark of ownership (Tomlin, this chapter). The remainder of the dishes was in coarse local fabrics, imitating BB1 (WHEAS 19, 5%; WHEAS 15, 16%; WHEAS 14.2, 3%). Sooting was noted on eight sherds from BB1 bowls and dishes, and one wheelmade Malvernian dish.

Mortaria, used in food preparation, made up a small proportion of the layer 1005 assemblage (Timby, this chapter), in common with that from the well at 1 The Butts (Fig. 5.22b). Few tankards were noted in any of the three groups (Fig. 5.22a–c), with none at all from the well at 1 The Butts (Evans, Chapter 3) though a near complete, 4th-century tankard was recovered during a watching brief on the same site (cf. Webster 1976, fig. 7, E44). This is worth noting, given that tankards have been the second most common Severn Valley ware form in other, earlier Worcester assemblages (Darlington and Evans 1992, 41; Bryant and Evans 2004, 247). Evidence from late Roman sites elsewhere hints that production of tankards may have declined during the 4th century. None were included in the assemblage from the late 4th-century Severn Valley ware kiln at Meole Brace, Shropshire (Evans *et al* 1999, 12), or from the contemporary Wroxeter, Tern-Severn production site (Houghton 1964), though both of these assemblages are small. Nor were any recovered from the late Roman villa site at Whitley Grange, Shropshire, not far from the Meole Brace kiln (Evans 2007). Closer to Worcester, at Malvern, tankards have been found at most of the known/ suspected production sites; at Newland Hopfields they represented 15–17% of the assemblage (Evans *et al* 2000, fig. 11, a and b). A notable exception to this is the late Roman, Hygienic Laundry kiln site (Peacock 1965–67; Evans *et al* 2000, table 13), which was producing a narrow range of jar forms similar to forms included in the various Butts assemblages.

Other vessels associated with the serving or consumption of liquids (flagons and beakers) were present in equally low numbers. Flagons were mainly in Severn Valley ware or sandy oxidised ware. Severn Valley ware forms (not illustrated) included fragments from an open-mouthed flagon (cf. Evans 2000, fig. 19, F11) and a cupped mouth flagon. The sandy ware flagon is illustrated (Fig. 5.19, 36).

Body sherds indicated the presence of Oxfordshire flagons in red colour-coated ware and parchment ware. The handful of beaker fragments was mainly in Nene Valley ware and included the rim from a pentice-moulded beaker (cf. Howe *et al* 1980, fig. 5, 55–57).

Pottery sources

Table 5.10 and Fig. 5.24a and b show that, although wares from regional sources were most common, a significant proportion of the assemblage from layer 1005 comprised 'Traded wares', ie widely traded wares produced by major industries outside the local area. In contrast, assemblages from earlier Worcester sites are more heavily dominated by regionally produced wares: at Sidbury traded wares comprised only 13% (Darlington and Evans 1992, table 1), at Castle Street only 9% (Griffin 2002, table 2), and at Deansway a maximum of 9% in Period 5 (Bryant and Evans 2004, table 49). The late Roman assemblages from the well fill at 1 The Butts (Fig. 5.24c) and the Castle Street evaluation (Dalwood *et al* 1997, appendix 4) show an even greater emphasis on traded wares.

This significant increase in traded wares indicates a change in the pattern of pottery supply to Worcester during the 4th century. This increased reliance on the major industries is consistent with the evidence from northern Britain, where similar changes seem to have occurred around *c* AD 350–55 (Evans 1999, 40). The small assemblage from the dark earth deposits at 8–12 The Butts does not fit with this pattern, though there are various biases there that may be distorting the proportions. Thus, residual sherds included heavy amphorae and a number of samian ware rims, exaggerating the proportion of imports. Residual amphora also appear to be one of the more common fabrics in the layer 1005 assemblage from 14–24 The Butts, at least when quantified by weight (Fig. 5.25a). Regional wares at 8–12 The Butts also included heavy, crudely formed sherds in handmade Malvernian ware, from large jars and bee hives or bread ovens; all paralleled in other 3rd- and 4th-century contexts (Darlington and Evans 1992, fig. 35, 4a, c; Peacock 1965–67, fig. 4, 81). These were also noted at 14–24 The Butts, but were not included in the pottery quantifications.

Figs 5.16 and 5.25a–b show that Severn Valley ware, though still the most common ware in the layer 1005 assemblage, is relatively less common here than on earlier Roman Worcester sites like Sidbury (Darlington and Evans 1992, table 1: 70% by count, 64% by weight), and Deansway (Bryant and Evans 2004, table 48: 66% by count in period 4). The well fill at 1 The Butts and the Castle Street assemblages (Dalwood *et al* 1997, appendix 4) produced even lower quantities of Severn Valley ware. Given that Worcester is in the heartland of Severn Valley ware production, this does seem to support the suggestion that the industry was in decline in the 4th century (Webster 1976, 44).

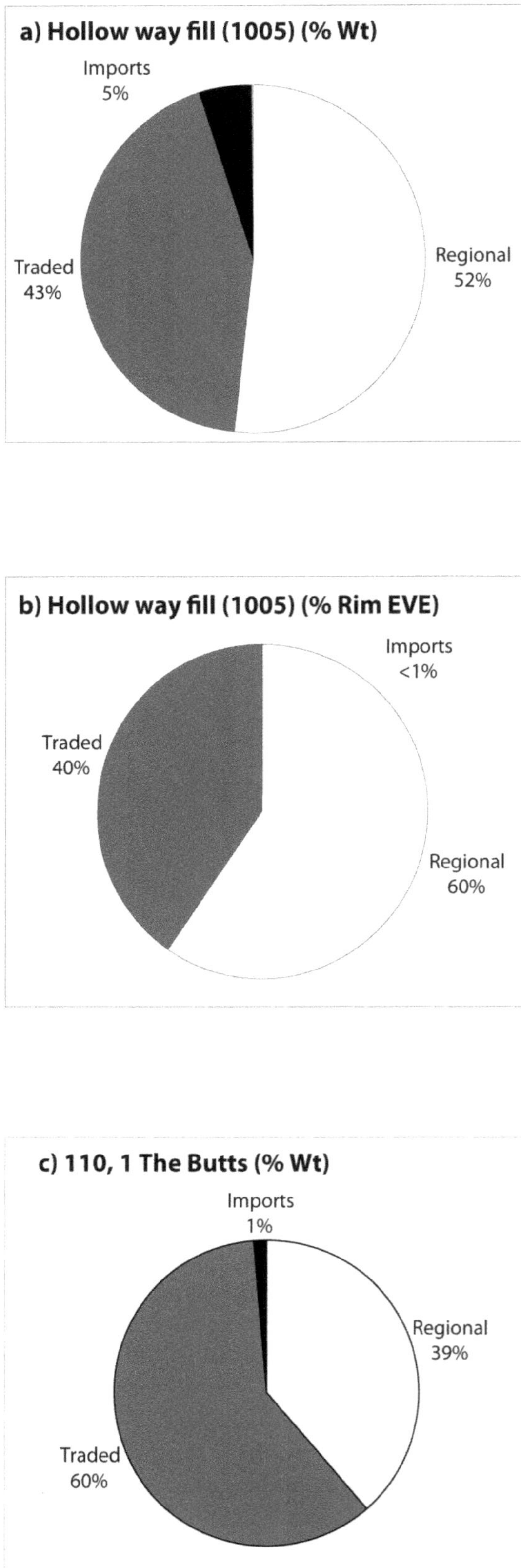

FIGURE 5.24 POTTERY FABRIC SOURCES

The relative proportion of BB1 cooking vessels, from Dorset, increased in 4th century Worcester. It was the second most common fabric in the layer 1005 assemblage, representing a much higher proportion than at Sidbury (Darlington and Evans 1992, table 1: 12% by count, 8% by weight) and Deansway (Bryant and Evans 2004,

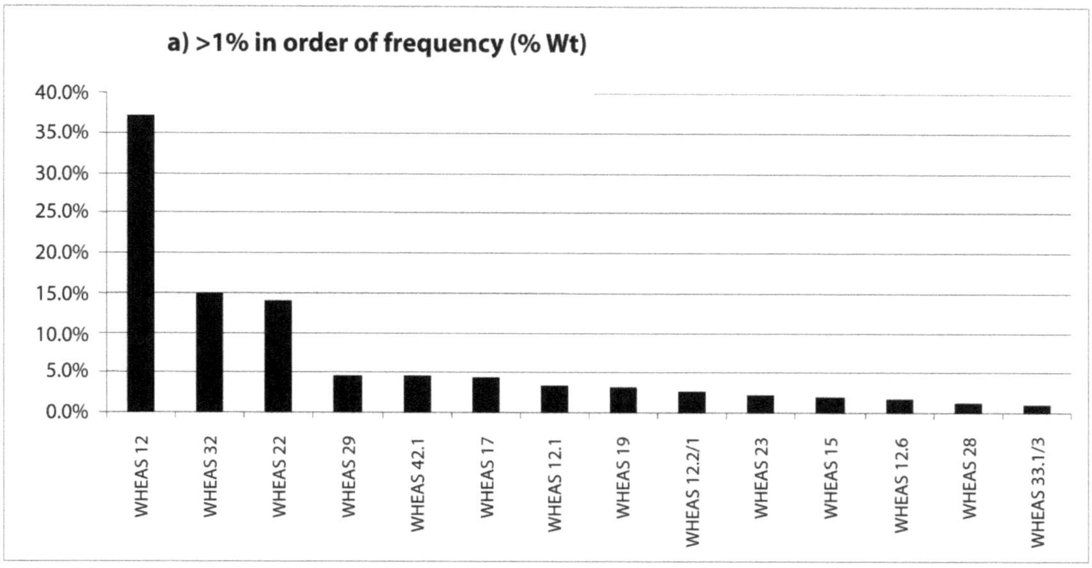

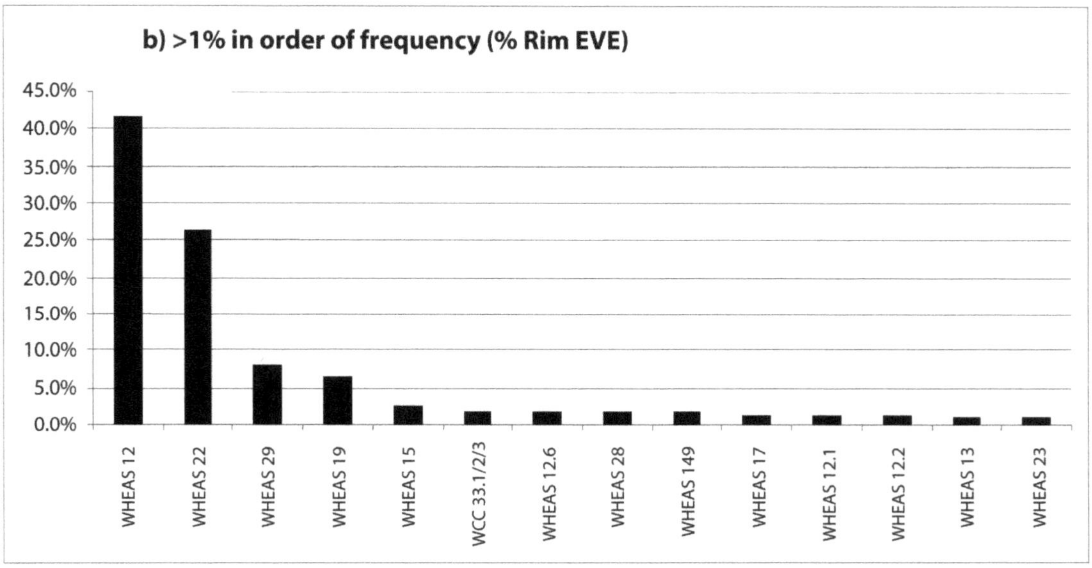

FIGURE 5.25 LATE ROMAN POTTERY FABRICS FROM LAYER 1005

table 48: 4% by count in Period 4). A similar increase is evident in the well at 1 The Butts and Castle Street assemblages, though not to the same degree Dalwood *et al* 1997, appendix 4). The data from 8–12 The Butts again seems ambiguous. An increase in BB1 has been noted on a number of 4th-century sites around the region: at Wroxeter (Symonds 1997, figs 370–71, 36%), Whitley Grange villa, near Wroxeter (Evans 2007, 55%) and Pentrehyling fort, Brompton, Shropshire (27%; Evans 2007, fig. 5.8p). High proportions of BB1 are also typical of later 4th- and 5th-century sites in North Wales (Evans 1999, 40).

The third most common fabric in layer 1005, though occurring in much smaller quantities, was Oxfordshire red colour-coated ware (Fig. 5.25a–b). This represented less than 1% by count at Sidbury and Deansway (Darlington and Evans 1992, table 1; Bryant and Evans 2004, table 49, period 5). The evidence from all but 8–12 The Butts, where it was entirely absent, indicates an increase in the supply of these colour-coated table wares during the 4th century. This is consistent with evidence elsewhere. In Warwickshire, for example, a relatively high proportion of 'fine' wares, including Oxfordshire products, has been noted on a number of 4th-century sites (Booth 1991, 1), and Oxfordshire wares are typical of late deposits at Kenchester (Tomber 1985, 126, table 2) and Wroxeter (Symonds 1997, 318). The proportion of this high status ware in the well deposit at 1 The Butts is particularly high, and may reflect the status, as well as the date, of the user of the pottery there. The associated demolition debris also hints at a building of some status nearby. Other Oxfordshire products, occurring in smaller quantities, included parchment ware (WHEAS 40) and white ware (WHEAS 38). Very small quantities of New Forest ware (WHEAS fabric 115) also reached the site, probably representing a single beaker.

Mortaria (Timby, this chapter) from the Oxfordshire kilns were also present in layer 1005, including white ware (WHEAS 33.1), white slipped ware (WHEAS 33.2) and red colour-coated ware (WHEAS 33.3). The other main source of mortaria was Mancetter-Hartshill (WHEAS 32). This was by far the most common source by weight (Fig. 5.25a), though Oxfordshire products were more common by rim EVE (Fig. 5.25b; WHEAS 32 not shown as <1%). All the mortaria, therefore, came from two major, widely traded industries. A 4th-century dominance of Oxfordshire mortaria over Mancetter Hartshill mortaria was noted at Bays Meadow, Droitwich. There, roughly two thirds of the late mortaria came from Oxfordshire, even though the Mancetter Hartshill kilns were much nearer (Barfield and Hartley 2006, 156).

Other relatively common wares in layer 1005 include the diagnostically late pink grog- (WHEAS 17) and shell-tempered wares (WHEAS 23), reduced Severn Valley ware (WHEAS 12.1) and wheelmade Malvernian ware (WHEAS 19). The proportions of these varied, sometimes dramatically, between the various Butts sites. Pink grog-tempered ware was very common at Castle Street, but absent from the excavated well deposits at 1 The Butts and the dark earth at 8–12 The Butts. Shell-tempered ware occurred in similar proportions at 14–24 The Butts and 1 The Butts, was by far the most common fabric at Castle Street and absent at 8–12 The Butts. Once again, this could reflect functional differences between these assemblages, or more subtle chronological differences. The remaining wares are all probably fairly local products, including Severn Valley ware variants, oxidised and reduced sandy wares. The latter were particularly common in the well deposit at 1 The Butts but occurred in smaller quantities on the other sites.

Common forms

The forms included in the hollow way (1005) pottery assemblage have been illustrated and described in detail above (Figs 5.18–5.20). The diagnostically late forms have been summarised within the dating discussion, and the functional range of forms has also been summarised and discussed above. In order to characterise the assemblage it is useful to also summarise the most common forms present to provide useful comparative data for future studies. This summary is illustrated graphically in Fig. 5.26, which plots the frequency of the different forms by % rim EVE.

Some of the chronologically diagnostic forms are amongst the more common types represented: eg, BB1 forms DD16.01, BI8.25, JK7.4 (Fig. 5.20, 52–54, 57–58); Oxfordshire red colour-coated ware bowl BC2.02 (Fig. 5.20, 65); Malvernian ware jar JK22.04 (Fig. 5.19, 46–47); sandy ware dish DA8.32 (Fig. 5.19, 40) and Severn Valley ware forms JW7.12 and BH17.30 (Fig. 5.18, 14–15, 19). It should be noted that the proportion of form DD16.01 may not be directly comparable with the others; this is an oval 'fish' dish (Fig. 5.20, 58) and the rim percentage had to be estimated. Other diagnostically late forms were less well represented, for example: Oxfordshire parchment ware bowl BA8.6 (Fig. 5.20, 73); Oxfordshire red colour-coated

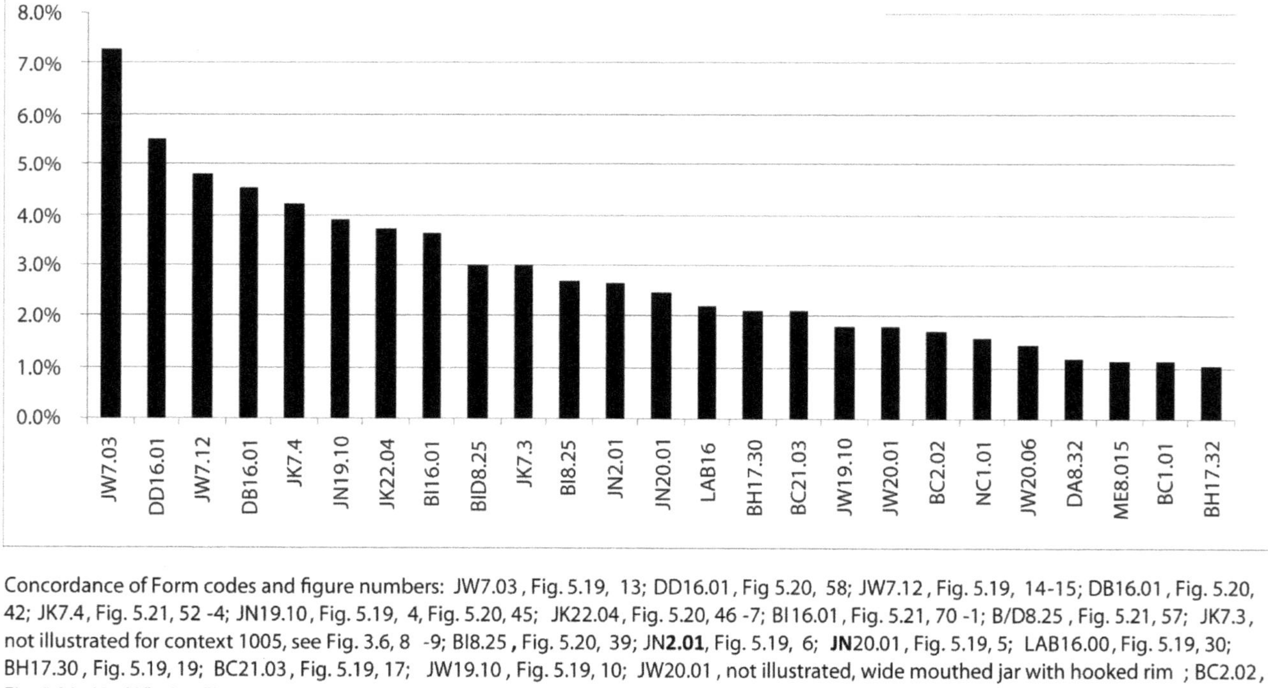

Concordance of Form codes and figure numbers: JW7.03, Fig. 5.19, 13; DD16.01, Fig 5.20, 58; JW7.12, Fig. 5.19, 14-15; DB16.01, Fig. 5.20, 42; JK7.4, Fig. 5.21, 52 -4; JN19.10, Fig. 5.19, 4, Fig. 5.20, 45; JK22.04, Fig. 5.20, 46 -7; BI16.01, Fig. 5.21, 70 -1; B/D8.25, Fig. 5.21, 57; JK7.3, not illustrated for context 1005, see Fig. 3.6, 8 -9; BI8.25, Fig. 5.20, 39; JN**2.01**, Fig. 5.19, 6; **JN**20.01, Fig. 5.19, 5; LAB16.00, Fig. 5.19, 30; BH17.30, Fig. 5.19, 19; BC21.03, Fig. 5.19, 17; JW19.10, Fig. 5.19, 10; JW20.01, not illustrated, wide mouthed jar with hooked rim; BC2.02, Fig. 5.21, 65; NC1.01, Fig. 5.19, 3; JW20.06, Fig. 5.19, 11; DA8.32, Fig. 5.20, 40; ME8.015, Fig. 5.21, 75; BC1.01, Fig. 5.19, 24, Fig. 5.20, 33, Fig. 5.21, 66; BH17.32, Fig. 5.19, 18

Figure 5.26 Hollow way (1005) pottery forms >1% in order of frequency (% rim EVE)

bowl BE7.02 (Fig. 5.20, 68–69); and the shell-tempered jars (JK7.02, JK20.01 and JK20.05 (Fig. 5.20, 59–61). A number of the more common forms were not closely datable, appearing to have been used over a long period of time. Amongst these are various Severn Valley ware jars, including the single most common type JW7.03 (Fig. 5.18, 13) and BB1-type plain-rimmed dish DB16.01 (Fig. 5.19, 42). Details of all these most common form types are included in the illustration catalogue above.

Pottery from Evaluation Trench 1

Another small group of late Roman pottery came from evaluation Trench 1, particularly from the truncated ditch at the southern edge of the trench (F104E). This produced a number of substantial, well-preserved sherds, including a variety of forms dating broadly to the late 3rd to 4th century, many paralleled in the assemblage from the hollow way (1005). Forms in BB1 (WHEAS 22) included conical bowls with dropped flange rims (Seager Smith and Davies 1993, type 25; cf. fig. 5.26, 57), cook pots with markedly splayed rims (*ibid*, type 3; cf. fig. 5.26, 52) and obtuse cross hatch burnish, a fish dish (*ibid*, type 21; cf. fig. 5.26, 58), and less diagnostic plain-rimmed dishes (*ibid*, type 20). Other diagnostically late Roman forms included Severn Valley ware jars (Webster 1976, fig. 6, 31; cf. fig. 5.24, 11), body sherds of Oxfordshire red colour-coated mortaria (WHEAS fabric 33.3), and an Oxfordshire parchment ware (WHEAS 40) carinated bowl (Young 2000, P24, fig. 27, 87; cf. fig. 5.26, 73). The latter was produced from 240–400+ but was particularly popular in the 4th century. A flat-hooped penannular brooch dating to the 4th century supported the pottery evidence (not illustrated, Nogueira, this chapter, 1005E). The absence of late Roman shelly ware (WHEAS fabric 23) might suggest that this assemblage is slightly earlier than the hollow way assemblage, or could simply be a product of the small assemblage size.

Layer 1002E, the upper fill of pit F100E (1003E) and the fill of pit F108E (1017E) all produced smaller, less diagnostic groups, which appeared to be broadly contemporary.

CONCLUSION

The most significant aspect of the Roman pottery assemblage from 14–24 The Butts is the late Roman material from the hollow way (1005), which provides the first substantial body of 4th-century pottery from Roman Worcester. This is complemented by the significant evidence for late Roman activity from other nearby sites in The Butts (Chapters 3 and 4). The combined evidence from these investigations adds considerably to the limited evidence for late Roman activity previously recorded from Worcester (Dalwood *et al* 1997).

Detailed analysis has provided an opportunity to study late Roman patterns of pottery supply and use. It has demonstrated an increasing reliance on the major national industries, particularly cooking vessels from southeast Dorset and table wares and mortaria from the Oxfordshire kilns. The evidence also points to a possible decline in the production and use of Severn Valley ware. The assemblage includes a number of diagnostically late Roman fabrics and forms, providing very good dating for the deposit, and associated dating for a number of less well dated forms. It thus enhances the dating of a number of regional forms, and provides further supportive evidence for the chronology of more widely traded wares. Analysis of the data has also allowed functional patterns to be assessed, based on vessel forms.

The excavations also produced a good quantity of 2nd- to early 3rd-century pottery which has also been studied in detail. The data for this has been compared with a number of larger, published assemblages from Worcester, particularly from Sidbury (Darlington and Evans 1992) and Deansway (Bryant and Evans 2004).

The detailed analysis and quantification of all this pottery adds to the growing body of data from Worcester. It has been recorded using the Worcestershire County fabric series, facilitating comparative studies. It will therefore make an important contribution to synthetic, perhaps spatial, analyses of pottery use and deposition in the Roman town. In particular, it fills a major gap with data for late Roman Worcester.

MORTARIA *Jane Timby*

The excavations produced 39 sherds of mortaria of which eight sherds are from Mancetter-Hartshill kilns, Warwickshire, one from an unknown Midlands source and the remainder from the Oxfordshire industries. The earliest piece is from F200 (2004), typologically datable to the 2nd century. The remaining examples all date to the 3rd or 4th century. Fabric codes after Tomber and Dore (1998).

Phased catalogue

Phase 1

1 Rimsherd. Orange buff exterior, pink core with white margins. A hard, granular, sandy fabric, possibly with quartz trituration grits. The fabric is one known from the West Midlands area and is known to occur in Warwickshire (Paul Booth pers. comm.) but is of unknown provenance. F200, 2004.

2 One bodysherd MAH WH. F304, 3004.

Phase 2

3 Oxfordshire whiteware mortaria (OXF WH), Young (2000) type M22. AD 240–400+. F104, 1012.

Phase 3

4 Sixteen bodysherds and two rimsherds OXF WH. Young (*ibid*) types M17 and M22. AD 240–400+. Area A, hollow way layer 1005.

5 Six bodysherds and two rimsherds Oxfordshire colour-coated ware mortaria (OXF RS). Young (*ibid*) type C97. AD 240–400. Area A, hollow way layer 1005.

6 One rimsherd Oxfordshire white-slipped mortarium (OXF WS), Young (*ibid*) type WC 4. AD 240–300. Area A, hollow way layer 1005.

7 Two bodysherds and five rimsherds from Mancetter-Hartshill mortaria (MAH WH). Minimum five vessels all with hammer rims and red-painted decoration on the flanges. 3rd century. Area A, hollow way layer 1005.

8 One bodysherd OXF RS. U/ S. AD 240–400.

Amphorae *D F Williams*

Seven amphorae bodysherds were recovered: three from the Area A hollow way layer 1005, Phase 3 (664g); one from Area E, F400, (4005), Phase 2 (20g); one from Area E, F404, (4012), Phase 1 (66g); and two from Area E, F405, (4013), Phase 1 (120g). All seven sherds recovered from the site belong to the commonly found globular-shaped amphora form Dressel 20, which carried olive oil from Baetica in Southern Spain and is found on many sites in Roman Britain, showing the wide distribution of olive oil in the Roman Empire (Williams and Peacock 1983; Peacock and Williams 1986, Class 25). The production of Dressel 20 was carried out over a long period, from the reign of Claudius until shortly after the middle of the 3rd century (Rodriguez-Almeida, 1989). Baetican olive oil was still exported after this date, though on a much reduced scale and in a smaller, thinner-walled version of Dressel 20 known as Dressel 23 (Carreras and Williams 2003). This long period of production and the lack of diagnostic features for the Worcester material mean that it is extremely difficult to accurately date these sherds within this time frame.

Samian ware *Felicity Wild*

Summary

The site produced only a small quantity of samian ware: 44 sherds, from about 36 vessels, mostly in small pieces. Eight sherds showed decoration, of which the four diagnostic sherds are described below. There was one potter's stamp. The material was all Central Gaulish and of 2nd-century date. Excluding scraps of uncertain form, forms were as follows: 37 (8 examples), beaker, probably 72 (1), 33 (1), 18/ 31 (1), 18/ 31 or 31 (3), 31 (3), 18/ 31R or 31R (2), 38 (1), 79 (1), 45 (1), cup (1), bowl (2), dish (2) (Table 5.11).

Although the assemblage is rather small, it is notable that the site produced no South Gaulish ware. The earliest material was likely to have been two sherds probably in the fabric of Les Martres-de-Veyre (decorated sherd no. 1 below and a scrap of uncertain form) dating to the first quarter of the 2nd century. Although the most closely datable pieces, the stamp and decorated ware, are Hadrianic or early to mid Antonine, the presence of the later Antonine plain forms (38, 79, 45) suggests that samian was arriving until the end of the 2nd century. The purely Antonine form 31 outnumbers the earlier form 18/ 31.

On a site predominantly of late Roman date, the samian is likely to be of limited use for dating purposes.

Area A

The upper fill of ditch F109 (1020), Phase 1, contained a sherd of beaker rim, probably form 72, dating to the 2nd century. The largest group of sherds came from the fill of the hollow way (1005), which contained three small decorated sherds, from different bowls, all of Hadrianic–Antonine or Antonine date.

Area B

Pits F200, F204 and F210 all produced samian of Antonine date. Form 31 from F200 (2004) and 79 from F203 (2007) are both datable to the second half of the 2nd century.

Area D

The material from cobble layer F300 (3001) was all of Hadrianic–early Antonine date. The upper fill of ditch F304 (3004) contained, as well as earlier material including no. 2 below, a sherd of the mortarium form 45, datable to *c* AD 170–200.

Area E

Most of the samian from Area E came from pits F404 and F405. The fills of F404 (4012, 4015, 4022) produced material dating to the second half of the 2nd century. The fills of F405 (4013) contained the stamp of Lottius ii, *c* AD 125–50 and a sherd probably in the style of Drusus i (no. 1, below), *c* AD 100–120.

Samian stamp

LOTTI.M Form 18/ 31–31. Die 3a of Lottius ii of Lezoux. The die has been recorded from forts in the Rhineland, which suggests activity before *c* AD 150. The potter may be the same as a man who worked at Les Martres-de-Veyre, and his activity at Lezoux is likely to be Hadrianic. *c* AD 125–50.

Decorated ware

In the following catalogue, figure types are quoted from Oswald 1936–37 (O), decorative motifs from Rogers 1974 (Rogers) and parallels from Stanfield and Simpson 1958 (S&S).

1 Form 37, Central Gaulish, probably in the fabric of Les Martres-de-Veyre. Small fragment showing an ovolo (Rogers B28), which was used with a similar

Context	Form	KILN	Date	Comment
BUW 03 1005	37	CG	Hadrianic–early Antonine	
	37	CG	Antonine	
	37	CG	Antonine	
	18/ 31 n 31	CG	Hadrianic–early Antonine	Rim fragment
	Dish (31 or similar)	CG	Antonine	Base fragment
	uncertain	CG	Probably late 2nd C	Heavy dish or bowl fragment
	18/ 31 n 31	CG	Probably Antonine	Burnt. Also another three fragments, probably of same
	scrap	CG (M de V?)	Trajanic–Hadrianic	
	scraps	CG	Hadrianic or Antonine	Two scraps (different)
1016	scrap	CG	Antonine	Abraded
1020	Beaker, probably 72	CG	2nd half of 2nd C	Rim fragment
	18/ 31R or 31 R	CG	Antonine	
Area A U/ S	37	CG	AD135–60	Cerialis II–Cinnamus style
2004	31	CG	Antonine	Also three small scraps, probably of same vesel
2007	79 or variant	CG	Late 2nd C	
2008	scrap	CG	Probably Antonine	
2014	scrap	CG	Probably Antonine	
3001	18/ 31R or 31R	CG	Hadrianic or Antonine	
	Prob 18/ 31	CG	Hadrianic–early Antonine	
	uncertain	CG	Hadrianic–Antonine	Base fragment
3004	45	CGF	*c* AD170–200	
	37	CG	Hadrianic	
	37	CG	Antonine	
	33	CG	Antonine	
	31	CG	Antonine	
4003	uncertain	CG	Antonine	scrap
4005	37	CG	Antonine	
4012	Bowl (37, 38 or 44)	CG	Antonine	2 fragments of rim
4013	18/ 31 - 31	CG	*c* AD125–50	Stamped LOTTI.M
	37	CG (prob M de V)	*c* AD100–120	Drusus style
	bow	CG	Hadrianic or Antonine	With rivet hole
4015	38	CG	Antonine (2nd half of 2nd century)	Two sherds, non-joining
4022	Flat-based dish	CG		

TABLE 5.11 SAMIAN WARE BY CONTEXT

bead row by X.3/ Drusus i. *c* AD 100–120. F405, Context 4013, Phase 1.

2 Form 37, Central Gaulish. Small fragment showing a tendril ending in a bud (probably Rogers G221), apparently springing from a festoon. Rogers G221 was used by the Potter of the Rosette, and what remains of the ovolo appears to fit that (Rogers B44) which he used with a wavy-line border, as here. Corner tendrils occur on work in his style (eg, S&S, plate 21, 261), though no examples are illustrated of them ending in G221, or springing from a festoon. The festoon looks like a poor impression of Rogers F40, used by X.13, or perhaps F41, and was not one of the Rosette Potter's normal repertoire of festoons. The Rosette potter worked on Les Martres-de-Veyre *c* AD 100–120. The fabric of the present piece does not appear to be that of Les Martres. A Hadrianic date seems probable. F304, context 3004, Phase 1.

3 Form 37, Central Gaulish, showing a warrior (O.177A) above a row of impressions of a triangular motif similar to, but slightly shorter than, Rogers U122. Both motifs and the same leaf-tip space filler occur on a bowl in a Cerialis ii-Cinamus ii style from Salzburg (Karnitsch 1971, taf. 38, 8), where the triangular motifs are used as oars in a boat and the 'warrior' appears to be fishing. The ovolo is likely to be the Cerialis-Cinnamus ovolo (Rogers B144) which also appears on the Salzburg bowl. *c* AD 135–60. Area A, U/ S.

4 Form 37, Central Gaulish. Lower edge of panel decoration, partially wiped off during the finishing of the bowl. The panel contains two impressions of a pointed leaf, either Rogers J56, used similarly at the base of panels by Casurius ii (eg, S&S, plate 134, 25, 28) or, perhaps more likely, the leaf-like motif Rogers U161, used by Cinnamus ii as well as Casurius and Do(v)eccus i. Of these, perhaps Cinnamus is the most likely, as Do(v)eccus and Casurius used larger beads than those in the beadrow here. Insufficient of the bud between the leaves remains for certain identification. The piece is likely to be Antonine. Area E, F400, 405.

GRAFFITI *Roger Tomlin*

Letters with a subscript dot are incomplete and the reading is not certain.

Area A, Layer 1005 (hollow way), Phase 3

1 Base sherd of an Oxfordshire red colour-coated ware dish. On the upper surface, within a circle incised while it was on the wheel, part of two letters incised after firing:
[…]**F#I, probably *[Ru]fi*, '(property) of Rufus'
H with unfinished cross-bar could be read the other way up, but the lower cross-bar of F is much more likely, followed by I. The lettering is unusually neat; indeed stylish capitals, with serif(s) to the I. *Rufus* is the only common name ending in *–fus* (Tomlin and Hassall 2006, 484, no. 57).

2 Area A, Layer (1005) (hollow way), Phase 3
Two conjoining base sherds of a BB1 'fish' dish. There is burnished decoration on the wall, and burnished strokes underneath, but none of them appear to be letters. Crudely incised underneath after firing, three lines meeting at angles, forming a triangle with one line extended:
P
Probably the initial letter of the owner's name, used as a mark of identification. Compare the next item.

3 Area A, U/ S
Base sherd of a black jar, scratched underneath after firing:
P̣
Only the upper stem and the loop (cut twice) survive, and the letter might be D, but this looks less likely. The sherd is too small to know whether the graffito was complete.

MEDIEVAL AND POST-MEDIEVAL POTTERY *Stephanie Rátkai*

INTRODUCTION

A small assemblage of post-Roman pottery consisting of 118 sherds, weighing 2855g, was recovered from 14–24 The Butts. The material was divided into medieval fabrics (following Hurst and Rees 1992) and post-medieval ware groups, and was quantified by sherd count and weight, and by minimum rim count. Vessel form was also recorded where identifiable. The fabrics and ware groups identified are listed chronologically in Table 5.12, along with their sherd counts by context. The amount of both medieval and post-medieval pottery is too small for detailed analysis, but some discussion of the results is worthwhile.

DISCUSSION

Medieval pottery

In total there were 37 medieval sherds, weighing 1,014g. The earliest post-Roman pottery was represented by a Worcester sandy glazed ware jug sherd (Worcester fabric F64.1). This dates to the later 12th or first half of the 13th century, and was recovered from the hollow way fill (1005). The remaining medieval pottery consisted almost entirely of Malvernian ware (Worcester fabric F69) which dates to the 14th to 16th century. There were few Malvernian form sherds, other than three bowls hollow way layer 1005, Phase 3; F400, 4005, Phase 2; F417, 4014, Phase 3; and a jar with a thumbed neck cordon (ditch F417, 4014). Most of the Malvernian sherds were glazed and it is likely that the greater proportion of these came from jugs. A small number of Cistercian ware cup sherds, dating to the late 15th to 16th century, were present in brick footings

	1002	1005	1011	1013	2014	2022	3002	4003	4005	4006	4014	4016	4017	4020	4025	4026	Total
Medieval fabrics																	
F64.1 Worcester glazed ware		1															**1**
F69 Malvernian ware		1			1			1	2		22	1		2			**30**
Cistercian ware							3								3		**6**
Post-medieval wares																	
Mottled ware		1			3	1							1				**6**
Mottled coarseware										1							**1**
Coarseware	5		3		1				1		4					1	**17**
North Devon gravel-tempered ware								1									**1**
Slip-coated ware			1			1		1									**3**
Feathered slipware				1				1									**2**
Trailed slipware								1									**1**
Brown salt-glazed stoneware	1			1													**2**
Light-bodied stoneware				1	1					1							**3**
Blackware		1		4	1	2											**8**
Creamware	8			1	3	3											**15**
Slipware																	**1**
Modern glazed wares	2			14		5											**21**
Post-medieval orange sandy ware						3											**3**
Miscellaneous																	
Roman oxidised wares									2		5						**7**
Total	**16**	**4**	**4**	**22**	**10**	**15**	**3**	**5**	**5**	**2**	**31**	**1**	**1**	**2**	**3**	**1**	**125**

TABLE 5.12 QUANTIFICATION BY SHERD COUNT OF ROMAN OXIDISED WARES AND POST-ROMAN POTTERY BY CONTEXT

(F301, 3002) and in the fill of a ditch (F415, 4025). In both cases the Cistercian ware sherds were the only ceramics in the contexts. Unfortunately, all the remaining medieval pottery was found residually.

The most striking feature of the medieval pottery is how little evidence there is of post-Roman occupation before the mid 14th century. The presence of only one early medieval sherd (the Worcester ware jug sherd) suggests that the absence of early medieval pottery genuinely reflects the absence of 12th- to 13th-century occupation here, and is consistent with garden use proposed by Kenyon (1999). More normal domestic occupation may have begun in the mid 14th or 15th century. The presence of Cistercian ware tends to suggest that there was definitely some domestic occupation in the vicinity from the late 15th or 16th century.

Post-medieval pottery

The absence of yellow ware amongst the post-medieval pottery is strongly suggestive of a second hiatus in occupation, although it may reflect clearance and truncation of 17th-century levels at the time of the Civil War. There is nothing amongst the post-medieval pottery which need be earlier than the late 17th or early 18th century. The period is consistent with the presence of mottled ware, North Devon gravel-tempered ware, slip-coated ware, feathered and trailed slipwares and both the light-bodied and brown-glazed stonewares.

Later 18th-century deposits were attested by the presence of creamware, and the sequence continued into the 19th century with industrial slipware, blue transfer-printed sherds, sponged ware and utilitarian whiteware etc (these fabrics are recorded as modern glazed wares in Table 5.12).

With the exception of a single sherd from Phase 3 hollow way layer 1005, the mottled ware sherds all had iron-rich fabrics which necessitated the use of a white underglaze slip. This makes it very unlikely that they are Staffordshire products, as buff firing clays are normally used there. Similar red-bodied mottled wares are known from the Welsh Marches (pers inspection by author) and in the post-medieval period there were clearly areas producing local imitations of the Staffordshire types. A sherd (recorded as mottled coarseware from feature F401 (4006), Phase 3) had a coarse, red, thick bodied 'coarseware fabric', but an internal and partial external white underglaze slip beneath a manganese mottled glaze.

Apart from the usual coarseware fabrics, there were three sherds from feature F218 (2022), Phase 4, with a very sandy orange fabric which contained abundant angular quartz *c* 0.01mm and sparse lumps of coarse red sandstone, up to 3mm. Two of these were body sherds with a glossy internal toffee brown glaze. The third sherd, a bowl rim, was similarly glazed but had two very thin horizontal white slip trails on the interior, just below the rim body junction. It is possible that all three sherds were from the same vessel. The sherds have been recorded as post-medieval orange sandy ware.

A single sherd of stratified post-medieval pottery was found during the evaluation (the remainder was in levels immediately beneath the tarmac in unstratified contexts). The sherd was a sgraffito slipware plate from Trench 2, 2006E. A layer of white slip was applied internally over an oxidised body. Sgraffito lines were then cut through to create a pattern and then green slip blobs and lines were applied to emphasise the pattern, part of which perhaps represented a fish. The finished design was then sealed beneath a clear lead glaze. No parallels have been found but the piece is certainly post-medieval in date and may be an import.

A small number of sherds was found in context 4009E in Trench 4. These included a substantial part of an unglazed red earthenware bowl or dish base, two sherds of black-glazed earthware, and a small rim sherd of modern glazed earthenware.

Worked stone *Rob Ixer and Erica Macey-Bracken*

Eighteen pieces of stone were recovered from four contexts at 14–24 The Butts hollow way layer 1005, Phase 3 x 6; F101, 1007, x 5; F304, 3005, Phase 1 x 2; F419, 4030 x 4; unstratified x 1. This material was quantified by count and weight and examined macroscopically.

Eleven of the 18 pieces of stone showed signs of having been worked. Five of these were sandstone blocks from a post-medieval wall (F101) in Area A, and four were from a 19th-century well in Area E. Thus only two pieces of worked stone are of possible Roman date. These were two fragments of tile from the hollow way fill (1005) in Area A. The smaller of the fragments was made from fine-grained, greyish-pink micaceous Lower Palaeozoic sandstone and had a nail hole. The larger fragment (catalogue no. 7) was a pinkish-grey, slightly micaceous Lower Palaeozoic sandstone and also had a nail hole measuring 6mm in diameter. Both of the stone types used for these tiles may be regional in origin.

Brick and tile *Erica Macey-Bracken*

All brick fragments recovered from the site were post-medieval or modern in date and are not reported on here. A total of 718 fragments of tile, weighing 41143g, were recovered from the evaluation and excavation. The fragments were generally small and unabraded, and no complete examples were recovered. The tile was quantified by count and weight, and scanned macroscopically to determine fabrics; eight different fabrics were identified. Their descriptions and their occurrence by context number are given in Table 5.13.

Examples of the known Roman tile forms *tegula, imbrex* and *tubulus* were all recovered from the site, as shown in Table 5.14. All the diagnostic sherds were either Fabric Two or Five. A further 104 undiagnostic fragments of Fabric Two, and a further 87 undiagnostic fragments were identified as Fabric Five, indicating that at least 31.9% of the tile recovered from the site was of Roman date. Medieval and post-medieval tile formed the largest part of the assemblage, although it is possible that some of the other undiagnostic fragments in other fabrics may also be of Roman date.

The medieval tile included glazed tile from the upper fill of Phase 3 ditch F417 (4014) and several nib tiles; the latter could be of post-medieval date.

Four tiles had nail holes (F202E, 2008E, Phase 2; F400, 4005, Phase 2; F404, 4012, Phase 1; F404, 4015, Phase 1. The holes were mainly square, apart from one example (4015), which was circular. The fragmentary nature of the tiles meant that it was not possible to judge the original position of the nail holes on the tile. Two tiles (F210, 2014, and 4003) were stamped with a capital T in a circle. Both of these tiles were from Phase 3 contexts, and are likely to be of post-medieval date. One tile (4003) had a possible graffito scratched into it, consisting of a deeply incised groove with a shallower J-shaped groove incised under it. On first impressions, this seems to be part of a crude drawing of an eye, but as the fragment is broken at this point, it is impossible to say for certain.

Fired Clay *Erica Macey-Bracken*

Amorphous lumps of fired clay were recovered from contexts attributed to Phase 1, Phase 2 and Phase 3. Most of the pieces are small and undiagnostic, although two fragments from Area E (F405, 4013, Phase 1) are flat on one side, as though they have been part of a surface. Two fragments from Area A (hollow way layer 1005, Phase 3) are of a very similar fabric to some of the tile, and may indeed be abraded pieces of tile.

Mortar *Erica Macey-Bracken*

One Phase 2 context (Area A, F100, 1011) yielded 15 fragments of mortar. Some of the fragments were very thick, and had one flat side, as if they had been part of a surface. These fragments had traces of whitewash on the flat surface. This suggests that the mortar was used to repair or even up a rough surface, possibly a wall (Steve Litherland, pers. comm.). One of the fragments has traces of 18th- to 19th-century brick adhering to it, which adds to the theory that this mortar is from a repair to a damaged wall, and suggests that these items represent contamination within the Phase 2 context.

Context	Fabric One	Fabric Two	Fabric Three	Fabric Four	Fabric Five	Fabric Six	Fabric Seven	Fabric Eight
Evaluation								
1002		2						
1003						1		
1004		4						
1005		4						
1015		1						
1017		4						
2003		5						
2004		3						
2006				1				
2008	8	3		3				
2013		1						
3003	1							
3005				3				
Trench 2 cleaning		1						
Excavation								
1004	3	1						
1005	4	86		1	61			
1011	9		1		5			
1012	1	1			2			
1013	1							
1014	1							
1018	7						1	
2004	3			1				
2007	1							
2008				1	1			
2010	3	1					1	
2013	2							
2014	18		1					1
2022	1							
2026	4							
3001	7							
3002	3							
3004	11	1			15			
3005	5				2			
3006					3			
3009	3				4			
4003	26		1	2				
4005	23		1		11		1	
4007	1							
4008				1	2			
4012	14				1			
4014	98		1	2		3		
4015	3							
4016	17	1	1	1	1			
4017	6			1				
4020	1							
4021	1				1			
4023/ 4	36							
4025	16			1				
4026	17			2				
4031	1							
Trench B Cleaning	1							
Trench B Southwest Corner	1							
U/S	8				2			

Tile fabric descriptions

Fabric One – Very hard, well-fired fabric. Dark orange-red throughout. Smooth external surfaces.

Fabric Two – Dark orange-red throughout, although a few examples have a grey core. Softer and sandier in appearance and touch to Fabric One.

Fabric Three – Very dense fabric, almost vitreous in appearance. Red-brown throughout.

Fabric Four – Coarse sandy orange-brown fabric with a dark grey core and occasional large subangular stone inclusions.

Fabric Five – Coarse, sandy orange-brown fabric with smooth, soft surfaces.

Fabric Six – Mid-grey sandy fabric with abundant grit inclusions.

Fabric Seven – Engineering brick fabric – hard-fired, dense purple fabric.

Fabric Eight – Pale orange-brown sandy fabric with smooth surfaces.

TABLE 5.13 FREQUENCY OF TILE FABRICS

Context	*Tegula*	*Imbrex*	*Tubulus*
1002E	1 (Fabric 2)	-	-
1004E Phase 2	2 (Fabric 2)	-	-
Trench 2020E cleaning	-	2 (Fabric 2)	-
1004	-	-	1 (1 x Fabric 2)
1005 Phase 3	14 (5 x Fabric 2, 9 x Fabric 5)	4 (3 x Fabric 2, 1 x Fabric 5)	4 (2 x Fabric 2, 2 x Fabric 5)
3004 Phase 1	3 (Fabric 5)	2 (1 x Fabric 2, 1 x Fabric 5)	1 (1 x Fabric 5)
3005 Phase 1	-	1 (1 x Fabric 2)	-
3006 Phase 1	3 (Fabric 5)	-	-
4016	-	1 (1 x Fabric 2)	-

TABLE 5.14 ROMAN TILE FORMS RECOVERED FROM 14–24 THE BUTTS

ENVIRONMENTAL EVIDENCE

MAMMAL AND BIRD BONES *Emma Hancox*

INTRODUCTION

Five boxes of bone (600mm x 400mm x 300mm) were recovered from the fieldwork. The vast majority of the assemblage came from 17th- to 19th-century contexts, which were too heavily contaminated to be considered for further work. The Roman assemblage is very small but, as it comes from the centre of Worcester, the results can be fed into the data already collected from other sites in the vicinity and it was therefore thought to be worth further study.

METHODOLOGY

The mammal bones were assessed following the standard protocol of Birmingham Zoological Laboratory, which is based on a modified version of the system described by Albarella and Davis (1994) and Davis (1992). This system considers certain anatomical elements as 'countable'; it does not include every bone fragment that is identifiable. The recordable skeletal elements considered are as follows: all the teeth, the skull (zygomaticus), scapula (glenoid articulation/ cavity), distal humerus, distal radius, proximal ulna, carpals 2–3, distal metacarpal, pelvis (ischial part of the acetabulum), distal femur, distal tibia, calcaneum (sustentaculum), astragalus (lateral part), naviculo-cuboid/ scafocuboid, distal metatarsal, proximal phalanges 1–3. At least 50% of the specified area has to be present to be 'countable'.

Horncores are considered as 'non-countable' elements using this system; however, they are recorded separately as noteworthy 'non-countables', along with any bones displaying evidence of butchery, pathology, burning or gnawing and any unusual species. These elements are not included in any statistical analysis.

The following skeletal elements were considered 'countable' for birds: scapula (articular end), proximal coracoid, distal humerus, proximal ulna, proximal carpometacarpus, distal femur, distal tibiotarsus and distal tarsometatarsus. All fish bone fragments were considered to be 'countable'.

Measurements were taken following von den Dreish (1976), Davis (1992) and Dobney *et al* (1996). The state of fusion was recorded as fused (F), fusing (G), ie spicules of bone have formed between the epiphysis and the diaphysis, but the line of fusion is still visible, or unfused (UM- unfused metaphysis/ diaphysis, UE – unfused Epiphysis, UX – both parts, unfused). Bird bones were recorded as fully ossified (F) or juvenile (UM), where complete ossification had not taken place.

Most bones were identified to species. The sheep/ goat distinction was only attempted on the humerus, metacarpal, tibia, astragalus, calcaneum, metatarsal, dP3, dP4 and horncore. Species identification on the galliform bones was only attempted on the scapula, coracoid, proximal ulna, carpometacarpus, proximal femur, distal tibia, and the proximal tarsometatarsus.

The upper and lower teeth of all mammals were recorded, with measurements taken on the lower teeth of equids, bovids, suvids and caprines and wear stages on the lower teeth of bovids, suvids and caprines. Mandibular fragments were considered to be 'ageable' when there were two or more teeth present with recognisable wear. Any abnormalities were noted, ie missing alveoli. Erupting teeth were coded following Grant (1982, 92–95): 'C – perforation in the crypt, V – tooth visible below the head of the bone, E – tooth erupting, H – tooth half erupted, U – tooth almost at full height, but unworn'.

FACTUAL DATA

The animal bone derived from a number of layers and discrete features from across the site, consisting of postholes, pits and linear features. The full results of the analysis are presented in Table 5.15. All the material considered in this report derives from contexts which have been dated to the 2nd to 4th century (Phases 1 and 2), as follows:

Context	Date	Preservation	Countable bones/ teeth						Ageable mandibles			Measureable bones/ teeth					Comments
			Bovid	Ovid	Suvid	Other	Bird	Fish	Bovid	Ovid	Suvid	Bovid	Ovid	Suvid	Other	Bird	
1002	19th c.	Fair		2									1				Saw and cut marks
1003		Fair	1									1					
1004		Fair	4									2					Chop and cut marks, gnawing
1005	17th–18th c.	Fair	136	43	33	14	2		6	6	3	22	13	4			Saw, chop and cut marks. Calcified. Other=12 equid teeth, 1 hare, 1 rabbit. 3 cow and 7 sheep/ goat horncores
1011	18th c.	Fair	6	2	1	8						1					Chop marks, cow horncore, Other=hare
1012		Fair/ poor	2	4		1	2			1		1	2				Cut and chop marks, Other=equid
1013	19th c.	Fair	1	4						1			1				Cut marks, gnawing
1015		Poor															Cut marks
1017		Poor	9	4	1												
1020		Fair/ poor	4	2					1	1		1	1				
2004		Fair/ poor	5	4	1												Chop marks, calcified bone, horncore
2014	18th c.	Good	3	2		2				1		1	1				Chop marks, Other= two dog mandibles
2022	19th c.	Fair	1														
3004		Fair/ poor	10	2	2	2	1					3					Cut marks, Other=dog and hare
3005		Fair/ poor	3	3	2								1				Very juvenile pig bones
3006		Fair/ poor		3													Gnawing
3009		Fair/ poor	1		1							1					Cut and chop marks, juvenile suvid bones
4003	17th–18th c.	Fair	5	1		3	1	1				2	1		1		Other= two equid and one rabbit
4005	17th–18th c.	Very poor	4														
4008		Poor	1														
4012		Very poor	1			1											Other=equid
4013		Poor	2	1						1			1				
4014	17th–18th c.	Fair/ poor	8					1									Chop and cut marks, calcified bone
4015		Fair	3	1		12									4		Other=8 equid and 4 dog. NCs-2 residual EQ MPs, cut marks on equid MC, pathology on vertebrae
4017	17th–18th c.	Fair	1														
4022		Fair		1													
4023/ 24		Poor	1														
4025	15th–16th c.	Poor		1													
4026	17th–18th c.	Poor				1											Other=EQ MP
U/ S		Varied	1	1	1	2	2			1			1				Cut and chop marks, horncore, Other=EQ
Totals			213	81	42	46	8	2	7	12	3	35	23	4	5	0	

TABLE 5.15 QUANTIFICATION OF MAMMAL AND BIRD BONES

Area A

1015, F107 Phase 1, 2nd century
1017 – Phase 1, 2nd century
1020, F109 – Phase 1, 2nd century
1012 – Phase 2, late 3rd to 4th century
1003 – Phase 2, 4th century

Area B

2004, F200 – Phase 1, late 2nd century

Area D

3004, F304 – Phase 1, late 2nd century
3005, F304 – Phase 1, late 2nd century
3006, F305 – Phase 1, late 2nd century
3009, F303 – Phase 1, late 2nd century

Area E

4012, F404 – Phase 1, late 2nd century
4015, F404 – Phase 1, late 2nd century
4022, F404 – Phase 1, late 2nd century
4008, F400 – Phase 2, late 3rd to 4th century

All of the assemblage considered here was hand collected. This has repercussions on the type and ratio of taxa found in the assemblage; without systematic and extensive sieving of samples there tends to be a bias towards larger species and anatomical elements. The state of preservation was mostly fair/ poor, with poor cortical integrity (exfoliation of the outer layers of the bone surface). The majority of the assemblage was fragmented, with only some contexts producing complete or almost complete bones.

Estimating the amount of residuality in faunal assemblages is extremely difficult. Various methods have been used to attempt this (eg, see Dobney *et al* 1996; Dobney *et al* 1997). One of the most common methods used is to employ pottery residuality levels as a baseline, although there may be no direct correlation, as different categories of artefact usually have different depositional pathways (refer to Evans and Millett 1992; Tomber 1991). The level of residual pottery at The Butts was low in the Roman contexts (Evans, pers. comm.).

A very low level of gnawing was observed with just three contexts showing evidence of this. It is not uncommon for a third of British vertebrate assemblages to be gnawed, and The Butts material was well below that threshold. This suggests that there is no secondary deposition caused by scavenging animals. No intrusive burrowing animals were noted from the Roman contexts.

Overview

A total of 78 countable bones and teeth were noted from the Phase 1 contexts. Their frequency of occurrence is listed in Table 5.16. Cattle made up 43% of this (based on NISP), with sheep/ goat being 28%, equids 14%, dog 10% and pig 5%. No small mammal, fish or bird bones were recorded; this is probably due to the relatively poor condition of the assemblage. The teeth and bones showed relatively similar percentages to each other (Table 5.16) with the exception of pig, which only made up 2% of the bones but 9% of the teeth, and horse/ donkey which made up 19% of the bones but just 6% of the teeth. The only other species identified was a 'non-countable' distal ulna of an unidentified bird. The ulna was most similar to those of the cormorant family, but did not match any of the reference collection bones.

Only 15 countable elements were recovered from the Phase 2 contexts, of which nine belonged to cattle, four to sheep/ goat, one to a woodcock and one to a chicken/ pheasant/ guinea fowl. This is too small a number to be considered for analysis. Due to the fragmentary nature and small size of the assemblage, it is not possible to carry out any analysis on the size and morphology of the bones.

The ratios of different elements in Phase 1 were interesting. Only one countable pig bone was recovered, an unfused metacarpal, and there were few sheep bones, although there appeared to be a relatively even spread of elements. However, of the cattle bones recovered, all but one (a tibia in ditch F304 (3004), Phase 1) were foot bones (metapodials, phalanges, calcanea, astragali and a scafocubiod).

Pathology was noted in ditch F304 (3004), Phase 1 in the form of two fused equid vertebrae. Butchery was recorded in ten contexts, including very fine cut marks on the metacarpal of a horse/ donkey (pit, F404 (4015), Phase 1). Burning was noted in three contexts in the form of several small fragments of both blackened and calcified bone.

DISCUSSION

The faunal assemblage is very small and, as a 'stand-alone' assemblage, it would not warrant much interest. However, as it comes from the centre of the Roman city,

Species	Percentage of the bones	Percentage of the teeth
Cattle	45%	41%
Sheep/ Goat	26%	31%
Pig	2%	9%
Equids	19% (but MNI only 2)	6%
Dog	8%	13%

TABLE 5.16 FREQUENCY OF SPECIES WITHIN PHASE 1 CONTEXTS

it was decided to record and measure the assemblage fully in the hope that it could feed into other research on faunal assemblages within the city.

The Deansway excavations (Dalwood and Edwards 2004, 518) produced a Phase 4 (AD 120 to 240) assemblage comprising 1,566 identifiable bones and fragments roughly contemporary with the 2nd- and early 3rd-century contexts from Phase 1 of 14–24 The Butts. Of the hand collected bone at Deansway, 60% was cattle and large ungulate, 29% sheep and small ungulate and 10% pig. Amongst the sieved material were found 52% cattle, 35% sheep and 13% pig. Further identifications included a number of horse bones, one red deer metatarsal and bones from at least three dogs. The sieved samples also contained domestic fowl, a shrew, field vole, frog, and fish bones from herring, eel and tope.

The material from 14–24 The Butts shows a similar pattern with 43% cattle, 28% sheep and 5% pig. Cattle were clearly the most important animal being utilised on site, particularly given that one cow produces a much higher meat yield than one sheep or pig. However, the cattle bones may not relate to domestic butchery waste, as the presence of large numbers of bovine metapodials and foot bones can sometimes be associated with tanning, as the skins were often brought to the tanning site with the feet bones still attached; these were then discarded in rubbish pits close to the tanning site. There is not enough evidence from the faunal material to suggest that tanning was taking place near here as it is such a small assemblage, but the presence of 20 foot bones out of the 21 bones recorded suggests that it is a possibility.

It is probable that the percentage of pig in the assemblage would originally have been higher, more similar to the 9% showed by the teeth, rather than the 2% showed in the bone. There are several reasons why the teeth and bones might show a different pattern for pig. The fact that the assemblage was all hand-collected will probably have produced a bias in favour of larger animals; for example, post-cranial cattle bones were probably preferentially collected over smaller animals as they were easier to see. Teeth are less susceptible to this problem as there is not a great size difference between the teeth of the main species recovered, suggesting that they probably give a more realistic percentage of the assemblage. Also, pig bones tend not to survive as well as some other species, due to being greasy and more porous. In addition, as pigs tended to be used solely as a meat producing animal, large numbers of juvenile animals are usually found in assemblages, unlike other animals such as sheep, which were kept alive for several years for their wool before being slaughtered for food. Juvenile bones are smaller and more porous than adult bones and do not survive as well, hence the number of pig bones recovered may be expected to be lower. It is also believed that pig bones were preferentially taken by scavenging dogs as they are more porous and greasy. The number of pig in the post-cranial assemblage is probably artificially low for these reasons.

The horse/ donkey metacarpal in F404 (4015), Phase 1, with evidence of fine cut marks, is interesting as this suggests that the animal was deliberately skinned. This has been noted at other sites in Britain. Horse bones from The Green in the Forebridge area of Stafford (Murray 2001) showed clear evidence of having been dismembered and skinned. The remains of at least four individuals were recovered from a large pit dated to the mid 18th century and Murray suggests that these were probably working animals that had reached the end of their lives and had been skinned and defleshed before being disposed of. The fused and ossified vertebrae in the same context could indicate that the equid bones in this context relate to old animals that either died or were killed because they were no longer useful and used for meat.

OYSTER *Erica Macey-Bracken*

Six fragments of oyster shell were recovered from phase 3 contexts (Area A, hollow way layer 1005, Phase 3 x 2; Area A, F105, 1013, Phase 3 x 2; Area B, F210, 2014, Phase 3 x 2). One piece has a large sub-circular hole but it is not possible to say if this hole was made deliberately.

CHARRED AND WATERLOGGED PLANT REMAINS *Wendy Smith*

INTRODUCTION

Archaeobotanical analysis of charred and waterlogged plant remains from 14–24 The Butts aimed to retrieve evidence for the cultivation and use of economic plants, primarily of cereals, and for environmental conditions in or around the site. Twenty samples were processed for charred plant remains, five of which proved sufficiently rich to merit full analysis. All five of these samples come from Phase 1 pit and ditch deposits dated to the 2nd and 3rd centuries. Their feature and context numbers are listed in Table 5.17. In addition to these, the coring of a well in watching brief Plot 5 (F901) provided six samples of waterlogged plant remains which are also dated to the Roman period.

METHODS

Samples from the Phase 1 pit and ditch deposits were processed by water flotation. The flots were sieved to 500μm and the heavy residues (the material which does not float) were wet sieved to 1mm. Both the flots and heavy residues were air dried at room temperature. The Birmingham Archaeology environmental officer examined the heavy residues for charred plant remains, but only charcoal was observed. The flots were sorted for charred plant remains by the author using a low-power binocular microscope at magnifications between x6 and x20.

Samples from the well (F901) in Plot 5 of the watching brief were obtained by coring, since conventional excavation would have been too hazardous. Coring produced a total depth of 7m of sediment, below which the sediment became too fluid to remain in the core head. This 7m core

AREA	Area E	Area D	Area D	Area D1	Area E	
FEATURE NUMBER	**F405**	**F304**	**F304**	**F305**	**F405**	
CONTEXT NUMBER	**4013**	**3004**	**3005**	**3006**	**4027**	
SAMPLE NUMBER	**12**	**13**	**14**	**15**	**23**	
CONTEXT TYPE	**Terminal of Ditch**	**Upper Fill of Ditch**	**Lower Fill of Ditch**	**Pit**	**Terminal of Ditch**	
PERIOD	**Roman**	**Roman**	**Roman**	**Roman**	**Roman**	
SAMPLE VOLUME (L)	**10 L**	**10 L**	**10 L**	**10 L**	**10 L**	
FLOT VOLUME (ml)	**30 ml**	**20 ml**	**20 ml**	**35 ml**	**20 ml**	
SEEDS PER LITRE	**47.3**	**21.6**	**9.9**	**58.6**	**19**	
LATIN BINOMIAL						**COMMON NAME**
CEREAL GRAIN						
Hordeum sp.	1	-	-	-	-	hulled barley
Triticum spelta L.	-	-	-	1	-	spelt
Triticum sp.	-	-	-	1	1	wheat
Triticum sp. – tail grain	1	-	-	-	-	wheat
Indeterminate cereal	-	12	-	4	-	indeterminate cereal
Indeterminate cereal/ large POACEAE	37E	-	6E	18E	7E	indeterminate cereal/ large grass
CEREAL CHAFF						
Hordeum sp. – six-row rachis node	-	-	-	1	-	six-row barley
Hordeum sp. – rachis node	1	-	-	-	-	barley
cf. *Hordeum* sp. – rachis node	-	-	-	6	-	possible barley
Hordeum sp. – awn fragments	4	-	-	31	-	barley
Triticum dicoccum Schübl./ *spelta* L. – spikelet fork	3 (=6gb)	2 (=4gb)	-	3 (5gb + 1r)	1 (=1gb +1r)	emmer
Triticum dicoccum Schübl./ *spelta* L. – glume base	86	71	41	74	31	spelt
Triticum spelta L. – spikelet fork	2 (=2gb + 1r)	-	-	3 (=5gb + 1r)	2 (=4gb)	spelt
Triticum spelta L. – glume base	90	26	10	63	44	spelt
Triticum spelta L. – glume/ lemma fragments	-	-	9	28	25E	spelt
Triticum cf. *spelta* L. – glume/ lemma fragments	30E	-	-	5	-	possible spelt wheat
Triticum sp. – rachis node	-	-	-	88E	-	wheat
Triticum sp. – glume/ lemma fragments	-	16	-	-	-	wheat
Triticum sp. – awn fragments	4	3	-	6	2	wheat
Indeterminate cereal – rachis node	125E	50E	5	179E	35E	indeterminate cereal
Indeterminate cereal – basal rachis node	2	-	-	-	-	indeterminate cereal
Indeterminate cereal/ large POACEAE – rachis node	-	-	12E	-	-	indeterminate cereal/ large grass
Indeterminate cereal/ large POACEAE – rachilla	-	-	-	1	-	indeterminate cereal/ large grass
Indeterminate cereal/ large POACEAE – awn fragments	-	-	-	10	-	indeterminate cereal/ large grass
COLEOPTILE						
Cereal/ Large POACEAE – coleoptile	8	7	1	13	2E	indeterminate cereal/ large grass
EDIBLE PULSES						
Vicia faba L.	-	-	-	-	1	broad bean/ celtic bean/ horse bean
cf. *Vicia faba* L.	-	-	-	-	1	possible broad bean/ celtic bean/ horse bean
TREE/ SHRUB						
Corylus avellana L. – nutshell fragment	-	1	-	-	-	hazel
WEED/ WILD PLANTS						
cf. *Urtica ures L.*	-	-	-	1	-	possible small nettle
Chenopodium sp./ *Atriplex* sp.	2	-	-	-	-	goosefoot/ orache
Montia cf. *fontana* L.	-	1	-	-	-	blink

Polygonum spp./ *Rumex* spp./ *Carex* spp.- indeterminate	1	-	-	-	-	knotgrass/ dock/ sedge
Rumex spp.	-	-	2	-	-	dock
cf. *Aphanes arvensis* L.	1	-	-	-	-	possible parsley-piert
Vicia sp./ *Lathyrus* sp.	-	-	-	-	1	vetch/ tare/vetchling
Melilotus sp./ *Medicago* sp./ *Trifolium* sp.	-	-	-	2	-	meliliot/ medick/ clover
Prunella vulgaris L.	-	-	-	1	-	self-heal
cf. *Galium* sp.	-	-	-	-	1	possible bedstraw
Tripleurospermum inodorum (L.) Sch. Bip.	1	1	-	2	-	scentless mayweed
Eleocharis palustris (L.) Roem. & Schult./ *uniglumis* (Link) Schult.	2	-	-	-	-	common/ slender spike-rush
Carex sp.	-	-	-	-	1	sedge
Lolium spp.	5	-	3	1	2	rye-grass
Poa sp.	-	1	-	-	-	meadow-grass
Avena sp. – awn fragments	-	12	2	17	1	oat
Avena sp. – rachis	1	-	-	-	-	oat
Avena sp./ *Bromus* sp.	10	-	-	-	2	oat/ brome
Phleum sp.	-	-	-	1	1	cat's-tail
Bromus sp.	1	-	-	-	-	brome
POACEAE – indeterminate large caryopsis	2	-	-	3	8E	grass family
POACEAE – indeterminate medium caryopsis	27	-	-	11	12	grass family
POACEAE – indeterminate small caryopsis	5	2	8	1	2	grass family
Unidentified fungal bodies (not included in count)	-	(2)	(8)	-	-	unidentified fungal body
Unidentified stalk – fragment	-	-	-	1	-	unidentified stalk
Unidentified – calyx (? *Agrostemma githago* L.)	-	1	-	-	-	unidentified calyx (? corn cockle)
Unidentified	2	1	-	6	-	unidentified
Indeterminate	19	9	-	4	7	indeterminate
TOTAL IDENTIFICATIONS	**473**	**216**	**99**	**586**	**190**	

Key: NE = estimated minimum number of identifications and (N) = items not included in the total count of taxa identified from a sample.

TABLE 5.17 CHARRED PLANT REMAINS FROM ROMAN DEPOSITS

was divided into six samples representing material from the following depths: 0–1m; 1–2m; 2–4m; 4–5m; 5–6m; 6–7m. In order to keep enough material for insect analysis (see Smith and Tetlow, this chapter), each sample for plant macrofossils was limited to 500ml of sediment. These 500ml samples were each washed over a 0.3mm geological sieve, and all of the material retained by the sieve was then rapidly scanned to assess the quantity and quality of plant (and insect) remains. Rich samples were then fully sorted under a low power binocular microscope at magnifications between x10 and x15.

Identifications were made at magnifications between x6 and x50 and in comparison with the author's own modern seed collection and that of Lisa Moffett (English Heritage, Birmingham), as well as in consultation with standard keys (eg, Beijerick 1976). Nomenclature for the plant remains follows Stace (1997) for indigenous species and Zohary and Hopf (2000) for cultivated species. The traditional binomial system for the cereals has been used here, following Zohary and Hopf (2000, 28, table 3 and 65, table 5).

RESULTS: CHARRED PLANT REMAINS FROM PIT AND DITCH DEPOSITS

Table 5.17 lists all the taxa identified in all five samples from the pit and ditch deposits, and Figure 5.27 provides a breakdown of the types of plants recovered in each sample, and their relative proportions. All of the samples are dominated by cereal chaff (accounting for between 73–85% of all identifications made in the samples), but include small quantities of cereal grain and weed/ wild seeds.

The cereal crops recovered include hulled barley (*Hordeum* spp.) and spelt (*Triticum spelta*). Rachis of barley was extremely limited (a total of seven identifications, five of which are only classified as possible barley (cf. *Hordeum* spp.), so it is not possible to determine whether six-row or two-row barley was cultivated. Glume bases and complete spikelet forks of spelt were abundant in all five samples. Small quantities of cereal coleoptiles (sprouts) were recovered from all five samples.

A single hazel (*Corylus avellana*) nutshell fragment was recovered from ditch F304 (3004), Phase 1. Whether this was accidentally incorporated in wood fuel or the

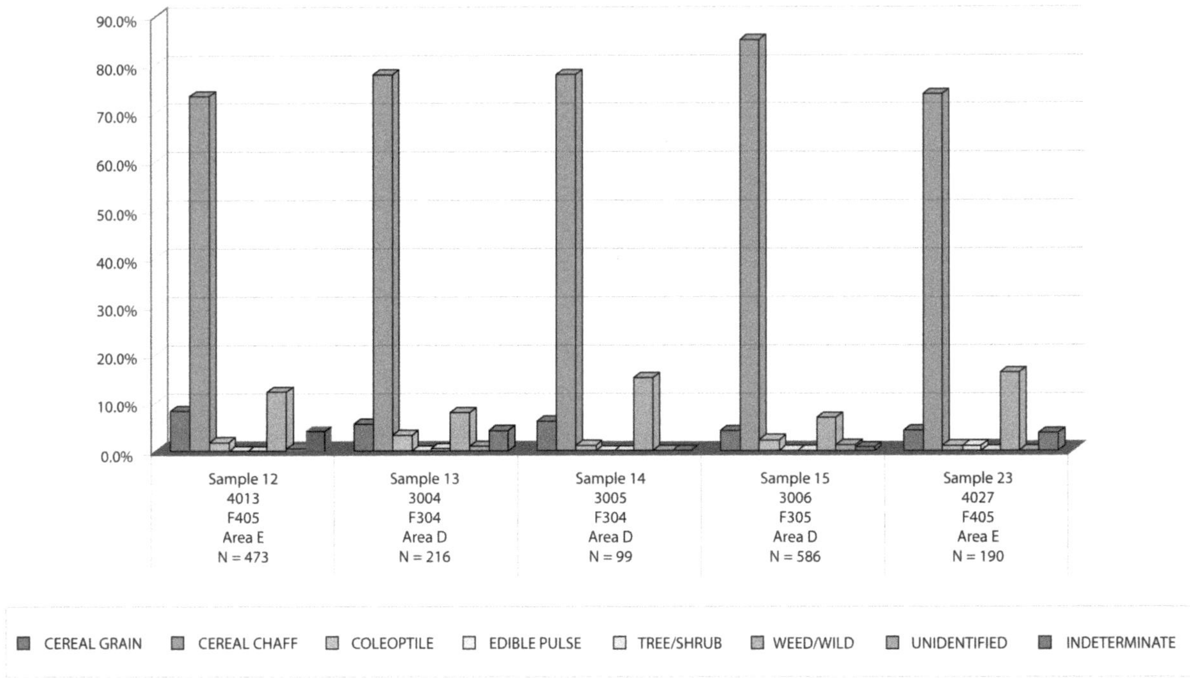

FIGURE 5.27 PROPORTIONS OF CHARRED PLANT REMAINS RECOVERED

remains of food is not clear. Certainly non-cereal crops and/ or collected wild foodstuffs are limited. The overall dominance of cereal crops in the samples studied here is unlikely to be due to any particular scarcity of non-cereal crops but, instead, reflects the pattern of charring events on site, which appear to frequently involve cereal crops.

The weed/ wild plants recovered from these samples most likely arrived in the deposits as weeds of the cereal crops, although it is possible that they could represent gardening waste or some other weeding debris (perhaps in field preparation). Modern agricultural practice, especially the use of herbicides, means that many of these taxa are rarely seen as weeds of cereal crops today; however, the weed/ wild flora recovered from 14–24 The Butts clearly contains species which are found frequently in association with ancient crop processing waste at other sites in Britain (eg, Greig 1990; Jones 1988a, 90; 1988b, 46).

Cereal crops in use

On the basis of these results, spelt (*Triticum spelta*) is clearly the main cereal in use, although small quantities of hulled barley (*Hordeum* spp.) grain and rachis nodes have also been identified. Remains of spelt are dominated by glume base and spikelet chaff fragments, with only one spelt grain securely identified for ditch F305 (3006), Phase 1.

Cultivation conditions

In most cases, only a few seeds were recovered, so the interpretations of soil conditions presented here must be treated with caution. Most of the weed/ wild plants recovered are not identifiable to species level; however, there is limited evidence from the weed flora for cultivation conditions. The recovery of sedge (*Carex* spp. – in pit F405, 4027, Phase 1) and common/ slender spike-rush (*Eleocharis palustris*/ *uniglumis* – in pit F405, 4013, Phase 1) most likely indicate damp, possibly wet conditions. Scentless mayweed (*Tripleurospermum inodorum* – recovered in pit F405, 4013, Phase 1; ditch F304, 3004, Phase 1; and pit F405, 4027, Phase 1) which typically occurs in waste, rough and cultivated land, is not particularly informative about soil conditions, although it is commonly associated with cereal crops in archaeobotanical assemblages. The recovery of self heal (*Prunella vulgaris*) in ditch F305, 3006, Phase 1, is also not particularly informative on soil conditions, as it is found in a range of soil types on grassland, woodland clearings and/ or rough ground.

Crop processing activities

The assemblages recovered from all five samples could be described as cereal crop processing by-products, which are primarily comprised of cereal chaff elements (ie, rachis nodes, glume bases and spikelets of spelt), with small quantities of cereal grain and similarly sized weed/ wild seeds. Although cereal chaff is less likely to survive charring than cereal grain (Boardman and Jones 1990), it clearly is abundant in these samples.

Spelt (*Triticum spelta*) is a hulled wheat which generally has two grains in each spikelet (individual segment) of the cereal ear. Although rarely grown today, hulled wheats do have a number of properties that would have been advantageous to past farmers. In particular, they can tolerate poor soil conditions and can resist a range of fungal diseases (Nesbitt and Samuel 1996, 42). During threshing, cereal ears of spelt will break up into individual spikelets, which contain grains surrounded by tough chaff. At this point the ancient farmer could either store or further process the spikelets of spelt. Storage of spelt (or other hulled wheats, such as emmer or einkorn) in spikelets is well known archaeobotanically and may serve to protect the grain from insect predation (Nesbitt and Samuel 1996, 52). In order to dehusk spelt to extract the grain, the spikelets must be pounded and the resulting mixture of freed grain and chaff is then winnowed, to separate light weed seeds and larger fragments of chaff from the grain, and then sieved, to remove any remaining weed seeds and smaller fragments of chaff from the grain (eg, Hillman 1984a; Jones 1984).

It is notable that all five samples have produced nearly identical assemblages, all of which are dominated by between 73% to 85% cereal chaff remains, most of which is spelt. This suggests that this by-product of cereal processing was repeatedly charred (most likely as a result of use of cereal chaff as fuel) and then disposed of into the ditches and pits located in this area. This material is likely to originate from two sources:

(1) It may be that by-products from sieving were intentionally destroyed or used as fuel. The combination of cereal chaff, with small quantities of cereal grain and weed seeds may represent 'chob' or 'cavings' (extraction by hand of weed seeds, cereal chaff, and tail grain during sieving – see Hillman 1984b, 2–3; Jones 1984, 46) which could be intentionally burned as fuel.

(2) It could be that this material represents accidentally burned crop processing material, perhaps in storage (as these by-products could be useful as feed for livestock) or even during small-scale crop processing (ie, cleaning small amounts of grain prior to hand-milling). However, the near total absence of grain, which is more likely to survive charring than cereal chaff (Boardman and Jones 1990), makes this interpretation fairly unlikely.

All of the samples are clearly secondary ditch and pit fills and, therefore, it is possible that much of the charred material has been reworked and/ or represents a mixture of a series of dumping events. As a result, any precise identification of the route(s) of arrival of the mixture of cereal chaff, cereal grain and weed seeds recovered from these deposits is not possible. Use of cereal processing waste as fuel is well attested (Hillman 1981, 1984a, 1984b) and disposal of spent fuel either into a ditch system or directly dumped on site, into pits for example, seems a likely explanation for the assemblages encountered.

Hedges and scrub

A single fragment of hazel (*Corylus avellana*) nutshell in Area D, ditch F304, 3004, Phase 1, provides limited evidence for the presence of hedgerows or scrub in the area. However, this could have entered the deposit in other ways, for example as a collected wild foodstuff or fuel waste. Hazelnuts, with their robust outer shells, however, are particularly suited to transport and storage, so although hazelnuts may have been locally available, they may also have been bought in or exchanged for some other product.

RESULTS: CHARRED AND WATERLOGGED PLANT REMAINS FROM THE WELL DEPOSIT

Rapid scanning of the six samples from the well core showed that only the three lower samples of sediment (4–7m depth) contained plant macrofossils in sufficient quantity for analysis. Table 5.18 presents the full list of charred and waterlogged taxa recovered from these three lowest samples of the well. In general, the results are very similar for all three samples and, certainly, they all are dominated by seeds of common nettle (*Urtica dioica*) and henbane (*Hyoscyamus niger*). It is notable, however, that Sample 4 (located at 4–5m below ancient land surface) was much less rich than the lower two samples. This, in combination with the near total absence of waterlogged plant remains and insect remains in higher samples (samples 1–3), may suggest that the upper part of the well was not preserved by waterlogging, or experienced periods of drying out, destroying any possible waterlogged remains.

Table 5.19 presents the range of habitats possible for those taxa identified to species level or those taxa not as precisely identified, but occurring in a limited in the range of habitats. They include taxa that are typical of grassland (possibly including pasture), waste/ open places and/ or cultivated (either arable or garden) ground. Some of the taxa recovered may also be indicative of woodland (possibly including woodland edge and/ or clearings).

The results from the three lowest samples of the well provide information about the nature of the surrounding environment of the feature, the nature of material deposited within the well once it fell out of use and its most likely date.

The environment surrounding the well

The plant macrofossils recovered from the lowest layers of the well are all highly typical of waste or neglected places (see especially Table 5.19). In particular, the presence of taxa such as bracken (*Pteridium aquilinum*), bramble/ blackberry (*Rubus* spp.), common nettle (*Urtica dioica*), small nettle (Urtica urens), common chickweed (*Stellaria media*), knotgrass (*Polygonum aviculare*) and henbane (*Hyoscyamus niger*), are all suggestive of areas of neglected and unmanaged ground. However, one cannot rule out the possibility that this area, located outside the walls of Roman Worcester, was used as pasture, fields or gardens. All of which could support many of the taxa listed in Tables 5.18 and 5.19, especially if the margins of the pasture/ fields/ gardens were fairly unkempt.

SAMPLE NUMBER **Depth** **Sample volume**	**Sample 4** **4–5m** **500 ml**	**Sample 5** **5–6m** **500 ml**	**Sample 6** **6–7m** **500 ml**	
Latin Binomial				**Common Name**
CHARRED PLANT REMAINS				
Triticum spelta L. – glume base	-	-	1	spelt
Triticum sp. – glume base	-	-	1	wheat
Triticum sp. – awn	-	-	1	wheat
Indeterminate cereal glume/ lemma fragment	-	+	1	indeterminate cereal
POACEAE – small caryopsis	-	2	-	grass – small-seeded
WATERLOGGED PLANT REMAINS				
cf. Cereal – bran fragment	-	-	1	possible bran fragment
Pteridium aquilinum (L.) Kuhn – leaves	-	+	++	bracken
Ranunculus acris L./ *repens* L./ *bulbosus* L.	-	2	4	meadow/ creeping/ bulbous
cf. *Ranunculus ficaria* L.	1	1	-	buttercup
Ranunculus subgenus BATRACHIUM (DC.)	1	2	1	lesser celandine
A. Gray	37	95	103	crowfoot
Urtica dioica L.	-	-	4	common nettle
cf. *Urtica dioica* L.	1	10	7	possible common nettle
Urtica urens L.	-	1	2	small nettle
Chenopodium murale L.	-	4	-	nettle-leaved goosefoot
Chenopodium sp.	-	-	1	goosefoot
Chenopodium sp./ *Atriplex* sp.	1	1	3	goosefoot/ orache
Atriplex sp.	-	7	17	orache
Stellaria media (L.) Vill.	1	-	-	common chickweed
cf. *Stellaria media* (L.) Vill.	-	3	14	possible common chickweed
Cerastium sp.	-	-	4	mouse-ear
cf. *Cerastium* sp.	-	1	-	possible mouse-ear
cf. *Silene* sp.	-	4	5	possible campion
Polygonum aviculare L.	1	-	-	knotgrass
Polygonum sp.	-	-	4	knotgrass
Polygonum sp./ *Carex* sp.	1	3	5	knotgrass/ sedge
Rumex spp.	-	3	4	dock
Malva sp.	-	1	-	mallow
BRASSICACEAE - indeterminate	1	1	1	cabbage family
Rubus spp.	2	4	5	bramble
Anthriscus sylvestris (L.) Hoffm.	-	2	-	cow parsley
cf. *Anthriscus sylvestris* (L.) Hoffm.	-	6	6	possible cow parsley
Torilis cf. *japonica* (Houtt.) DC.	-	-	1	upright hedge-parsley
APIACEAE - indeteminate				carrot family – indeterminate
Hyoscyamus niger L.	8	20	11	henbane
cf. *Hyoscyamus niger* L.	-	-	1	possible henbane
Stachys sp.	-	6	2	woundwort
cf. *Stachys* sp.	-	1	1	possible woundwort
Galeopsis sp.	-	1	-	hemp-nettle
Prunella vulgaris L.	-	1	-	selfheal
Plantago major L.	-	1	-	greater plantain
Euphrasia spp./ *Odontites* spp.	-	1	-	eyebright/ bartsia
Sambucus nigra L.	3	2	6	elder
Onopordum acanthium L.	1	-	1	cotton thistle
Picris sp.	-	1	-	oxtongue
Sonchus sp.	-	-	1	sow-thistle
ASTERACEAE – indeterminate	1	-	2	daisy family – indeterminate
cf. ASTERACEAE – indeterminate	-	-	1	possible daisy family –
Carex sp. – 3-sided	1	4	4	indeterminate
Juncus sp.	-	1	6	indeterminate sedge
cf. *Juncus* sp.	-	-	1	rush
POACEAE – indeterminate small caryopsis	5	10	18	possible rush
POACEAE – indet. small caryopsis/ indet.	1	-	-	grass family – indeterminate small-
Cereal – tail grain	1	-	2	seeded
POACEAE – indeterminate large caryopsis	2	-	1	small-seeded grass/ cereal tail grain
Unidentified broadleaf	1	1	1	grass family – indeterminate large-
Unidentified bud	7	-	1	seeded
Unidentified leaf abscission pad	++	+	+	unidentified broadleaf
Unidentified moss	1	2	-	unidentified bud
Unidentified thorn	3	31	20	unidentified lead abscission pad
Unidentified				unidentified moss
				unidentified thorn
				unidentified
TOTAL	**82**	**236**	**276**	

Key: + = < 5 fragments, ++ = 5 to 15 fragments and +++ = > 15 fragments

TABLE 5.18 CHARRED AND WATERLOGGED PLANT REMAINS FROM THE WELL

Latin Binomial	Arable fields	Cultivated ground	Damp places	Dry, acid soils	Grassland	heaths/ moors	Hedgebanks/ Hedgerows	Nitrogen-rich soils	Open (unshaded) ground	Rough ground	Water	Waste ground	Wet places	Woodland (incl. Woodland clearings)	*Common Name*
Pteridium aquilinum (L.) Kuhn – leaves	-	-	-	✓	-	✓	-	-	-	-	-	-	-	✓	bracken
Ranunculus subgenus RANUNCULUS	-	-	-	-	✓	-	-	-	-	-	-	-	✓	-	meadow/ creeping/ bulbous buttercup
cf. *Ranunculus ficaria* L.	-	-	✓	-	-	-	✓	-	-	-	-	-	-	✓	possible lesser celandine
Ranunculus subgenus BATRACHIUM (DC.) A. Gray	-	-	-	-	-	-	-	-	-	-	✓	-	-	-	crowfoot
Urtica dioica L.	-	✓	-	-	-	-	-	✓	-	-	-	-	✓	✓	common nettle
Urtica urens L.	-	✓	-	-	-	-	-	-	-	-	-	✓	-	-	small nettle
Chenopodium murale L.	-	✓	-	-	-	-	-	-	-	-	-	✓	-	-	nettle-leaved goosefoot
Stellaria media (L.) Vill.	-	✓	-	-	-	-	-	-	✓	-	-	-	-	-	common chickweed
Polygonum aviculare L.	-	-	-	-	-	-	-	-	✓	-	-	-	-	-	knotgrass
Anthriscus sylvestris (L.) Hoffm.	-	-	-	-	✓	-	✓	-	-	-	-	-	-	✓	cow parsley
Torilis cf. *japonica* (Houtt.) DC.	-	-	-	-	✓	-	✓	-	-	-	-	-	-	✓	upright hedge-parsley
Hyoscyamus niger L.	-	-	-	-	-	-	-	✓	-	✓	-	✓	-	-	henbane
Prunella vulgaris L.	-	-	-	-	✓	-	-	-	-	✓	-	-	-	✓	selfheal
Plantago major L.	-	✓	-	-	✓	-	-	✓	-	✓	-	-	-	-	greater plantain
Sambucus nigra L.	-	-	-	-	-	-	✓	-	-	✓	-	✓	-	✓	elder
Onopordum acanthium L.	✓	-	-	-	-	-	-	-	-	✓	-	✓	-	-	cotton thistle
Carex sp. – 3-sided	-	-	-	-	-	-	-	-	-	-	-	-	✓	-	indeterminate sedge
Juncus sp.	-	-	-	-	-	-	-	-	-	-	-	-	✓	-	indeterminate rush

Ranunculus subgenus RANUNCULUS includes *Ranunculus acris* L./ *repens* L./ *bulbosus* L.

TABLE 5.19 HABITAT RANGES OF SELECTED WATERLOGGED TAXA FROM THE WELL DEPOSIT

Material deposited in the well after it fell out of use

Charred material recovered from the well was limited to the lowest sample (sample 6, 6–7m below ancient land surface). A limited quantity of charred cereal remains, including spelt (*Triticum spelta*) and indeterminate wheat (*Triticum* spp.) glume bases were recovered. The sample size was extremely small (just 500 ml), but if this was factored up to the average sample size of 20L – this would result in 80 identifications of spelt or indeterminate wheat glume bases. This projection is not a particularly rich result, but would be consistent with small quantities of general crop processing detritus entering the well during its use.

Most plant remains recovered from this lowest sample are primarily waterlogged and those from the next two samples above this are entirely waterlogged. The majority of waterlogged plant macrofossils accumulated in the well is most likely indicative of waste ground conditions immediately around the well or possibly at the margins of areas in the vicinity of the well, such as pasture, gardens or fields. Whether the well was infilled slowly or rapidly is not clear; however, the uniformity of results from the bottom 3m of the well is notable. It is possible that this sort of material could have been collected in clearing/ weeding areas around the well and dumped down it, after the well fell out of use. However, it also is possible that this material represents the gradual accumulation of plant matter, possibly through collapses at the top of the well.

Evidence for the date of the use of the well

The recovery of spelt glume bases from the well provides some evidence for the date of its use. Spelt was introduced to the British Isles in the Late Bronze or Iron Age and fully adopted by the Roman period (Hillman 1981; Jones 1981). In some areas, spelt continued to be cultivated into the Saxon period (Hillman 1981; Jones 1981). In addition, the recovery of cotton thistle (*Onopordum acanthium*), which is prevalent in many Roman wells from the Midlands (eg, Greig 1988) and requires warmer growing conditions than exist today, also suggests that this deposit is Roman in date. Thus, it seems likely that the spelt and cotton thistle recovered from the lowest sample (sample 6) of the well were deposited during the Roman period, and that the well therefore fell out of use during this period. The fill of the cut for the well (9000) produced fragments from 2nd- and

3rd-century pottery vessels (Evans, this chapter). A single achene of cotton thistle was also recovered from Sample 4 (4–5m below ancient land surface) and twelve sherds of pottery within the later backfill were broadly dated to the late 3rd or 4th century.

COMPARISONS WITH OTHER SITES

Other Roman period sites in Worcester that have produced substantial charred plant remains include Deansway (Moffett 2004a), Farrier Street (De Rouffignac 1994), Sidbury Street (De Rouffignac 1992) and Blackfriars (Moffett 1987). Of these, Farrier Street, Sidbury Street and Blackfriars all report assemblages dominated by cereal chaff remains, primarily of spelt (*Triticum spelta*), which is consistent with the results from 14–24 The Butts. At Deansway more than half the samples were again rich in chaff or a mixture of chaff and weeds, but here there were also some grain-rich samples containing over 90% indeterminate *Triticum* spp. Grains. This seems to have been cleaned grain which had been more or less fully processed and made ready for consumption, whilst the chaff-rich samples from Deansway were more likely to have been used as tinder or fuel, as suggested here for 14–24 The Butts.

Latin Binomial	**Alcester (sample 168M)**	**The Butts (MR1) (samples 4–6)**	**Droitwich**	**Mancetter Well 17/ 23**	**Tiddington Well 579 (sample 10/ 1)**	**Tiddington Well 131 (sample 5/ 5)**	**Tiddington Well 127 (sample 7/ 1)**	***Common Name***
Pteridium aquilinum (L.) Kuhn – leaves	-	-	-	-		-	-	bracken
Ranunculus subgenus RANUNCULUS								meadow/ creeping/ bulbous buttercup
Ranunculus parviflorus L.		-	-	-				small-flowered buttercup
Ranunculus sceleratus L.	-	-		-	-	-	-	celery-leaved buttercup
cf. *Ranunculus ficaria* L.	-		-	-	-	-	-	possible lesser celandine
Ranunculus subgenus BATRACHIUM (DC.) A. Gray	-		-	-	-	-	-	crowfoot
Papaver dubium L.	-	-	-	-			-	long-headed poppy
Papaver argemone L.	-	-	-	-	-			prickly poppy
Fumaria sp.	-	-		-	-	-	-	fumitory
Urtica dioica L.		!	!			!		common nettle
cf. *Urtica dioica* L.	-		-	-	-	-	-	possible common nettle
Urtica urens L.	!				!			small nettle
Chenopodium rubrum L./ *chenopodioides* L.) Aellen	-	-		-	-	-	-	reed/ saltmarsh goosefoot
Chenopodium murale L.	-	-	-	-	-	-	-	nettle-leaved goosefoot
Chenopodium cf. *album* L.		-				!	-	fat-hen
Chenopodium sp.	-		-	-	-	-	-	goosefoot
Chenopodium sp./ *Atriplex* sp.	-		-	-	-	-	-	goosefoot/ orache
Atriplex sp.				-			-	orache
Montia fontana ssp. chondorosperma (Fenzl) Walters	-	-	-				-	blink
Arenaria sp.	-	-	-	-			!	sandwort
Stellaria media (L.) Vill.				!				common chickweed
cf. *Stellaria media* (L.) Vill.	-		-	-	-	-	-	possible common chickweed
Stellaria graminea L.		-	-	-			-	lesser stitchwort
Stellaria uliginosa Murray	-	-	-	-	-		-	bog stitchwort
Stellaria sp. (immature)	-	-	-	-		-	-	immature stitchwort

1 Key: (= Polygonum persicaria L.) – indicates change in nomenclature between Stace 1997 and earlier Floras. Shading indicates presence and ! indicates dominant taxa in the assemblage (more than one taxa may have an (!) – if no single taxon is clearly dominant). Source for Alcster, Droitwich, Mancetter and Tiddington data is Greig 1988.

2*Ranunculus* subgenus RANUNCULUS includes *Ranunculus acris* L./ *repens* L./ *bulbosus* L.

TABLE 5.20 COMPARISON OF WATERLOGGED PLANT REMAINS FROM THE WELL DEPOSIT WITH OTHER WELL DEPOSITS FROM THE MIDLANDS

At Blackfriars, a very rich deposit of burnt spelt chaff points to its possible use as a proper fuel, rather than just as tinder (Moffett 1987). Indeed, at Salford Priors in Warwickshire, where spelt glume bases and spikelet forks dominated 1st- to 4th-century samples, it was suggested that not only was spelt chaff used as fuel, but '[the] availability of such large quantities of chaff material suggests that spelt was being processed on a large scale, with the chaff being sorted as a product in its own right ready for use when needed' (Moffett and Ciaraldi 1999, 170). Several other sites in Warwickshire have also produced archaeobotanical assemblages dominated by cereal chaff (especially spelt) remains. At Billesley Manor, Warwickshire, (Monckton 1999a) all of the 3rd- to 4th-century samples were dominated by spelt and indeterminate emmer/ spelt chaff. At Glebe Farm, Bubbenhall (Monckton 1999b), five of the 12 Roman period samples studied (dating from the 2nd to 3rd century) were dominated by spelt and indeterminate emmer/ spelt chaff (glume bases and spikelet forks).

The question of whether sprouted grains represent malting or spoiling remains problematic in the absence of associated structures such as corn driers. Van der Veen (1989), in her review of Roman corn driers, suggests that sprouted grains or detached sprouts (coleoptiles) must appear in high percentages to be securely interpreted as malt, rather than a spoiled crop (ie, grain which has become damp and started to germinate). At Deansway (Moffett 2004a,b), more than half the grains had germinated in two of the ten Roman period samples, but as Moffett cautions, even sizeable caches of grain can become spoiled and might well be disposed of by burning. Coleoptiles could represent the 'cumins', or waste, from rubbing parched or malted grain, as suggested by Pelling (2000, 54–55) at the Hockley Chemical Works in Alcester. Here a chaff-rich sample from the lower fill of a pit contained 201 coleoptiles (sprouts). However, this represented only 5% of the identified remains in the sample, and it may simply be residual waste mixed in with fuel.

Perhaps the most significant aspect of the plant remains from 14–24 The Butts is the discovery of a waterlogged assemblage from a Roman well. Indeed, this appears to be the first recorded assemblage of waterlogged plant remains from Worcester. In the wider context of the Midlands, Roman well deposits have previously been investigated at Alcester, Droitwich, Mancetter and Tiddington (Greig 1988). Table 5.20 compares the results from 14–24 The Butts with these other Midland sites, presenting a full list of taxa recovered from all these sites updated to the nomenclature of Stace (1997). Interestingly, the results are quite similar at all three sites. Taxa such as common nettle (*Urtica dioica*), small nettle (*Urtica urens*), fat hen (*Chenopodium album*) or indeterminate goosefoots (*Chenopodium* spp.), knotgrass (*Polygonum aviculare*), henbane (*Hyoscyamus niger*) and rush (*Juncus* spp.) are all abundant in these samples (Greig 1988, 372). Two taxa, hemlock (*Conium maculatum*) and cotton thistle (*Onopordum acanthium*) appear to be much more prevalent in Roman times and both require warmer growing conditions than exist today (Greig 1988, 373). Hemlock was not recovered at The Butts, but the presence of cotton thistle may suggest a generally warmer climate in the area than at present.

Conclusion

The plant assemblage from the five samples from pit and ditch fills at 14–24 The Butts demonstrates not only the consistent use of spelt chaff, but also a consistent pattern of deposition. It is possible that cereal crop processing waste was used as fuel and then intentionally disposed of in ditches and pits, and that this represents a repeated pattern in the treatment of such cereal crop processing by-products at the site. Intentional dumping of charred debris (ie, spent fuel, spoiled crops, accidentally charred material, etc) seems the most likely explanation for the formation of the deposits encountered here.

The plant remains recovered from the well appear to be consistent with Roman well deposits from elsewhere in the Midlands (eg, Greig 1988). The recovery of charred spelt (*Triticum spelta*) and indeterminate wheat (*Triticum* sp.) glume bases, as well as a single waterlogged cotton thistle (*Onopordum acanthium*) achene, supports a Roman date for the deposition of material into the lowest sampled deposit (sample 6, 6–7m below ancient land surface) from this well. The date(s) of the upper samples, however, remains uncertain. Nevertheless, the initial infilling of the well appears to have commenced in the Roman period. The majority of taxa recovered from the three samples studied are typical of areas of waste ground. This could indicate that the area immediately around the well was overgrown or that unkempt, overgrown areas existed at the margins of fields, gardens and/ or pasture in the vicinity of the well.

Insect remains from the well *David Smith and Emma Tetlow*

Introduction

The 7m sediment core from the well (F901, Plot 5) was divided into six samples representing the following depths: 0–1m; 1–2m; 2–4m; 4–5m; 5–6m; 6–7m. These were labelled samples 1–6 respectively, and were examined for insect remains in order to retrieve information about the environment surrounding the well.

Sample processing and analysis

The soils samples were paraffin floated using the standard method as outlined in Kenward *et al* (1980). The weights and volumes of the samples processed are presented in Table 5.21.

The insect remains present were sorted from the paraffin flots using a low power binocular microscope and stored in ethanol. The Coleoptera (beetles) were identified by direct comparison to the Goraham and Girling collections of British Coleoptera. The various taxa of insects recovered

			No 4, 4–5m	No 5, 5–6m	No 6, 6–7m
Processed Weight					
Processed Volume					
COLEOPTERA					
Carabidae					
Bembidion spp.	oa		-	1	-
Harpalus aeneus (F.)	oa		-	1	-
Harpalus spp.	oa		1	-	1
Stenolophus skrimshiranus Steph.			-	-	1
Pterostichus melanarius (Ill.)	oa		-	-	1
Calathus fuscipes (Goeze.)	oa		4	-	-
Syntomus foveatus (Fourcr.)	oa		-	-	1
Hydraenidae					
Hydraena spp.	oa-w		-	1	1
Octhebius spp.	oa-w		1	-	1
Helophorus spp.	oa-w		1	-	-
Hydrophilidae					
Cercyon impressus (Sturm.)			1	-	-
Cercyon haemorrhoidalis (F.)	rf	sf	-	-	1
Cercyon analis (Payk.)	rt	sf	2	1	1
Histeridae					
Onthophilus striatus (Forst.)	rt	sf	-	-	1
Abraeus globosus (Hoffm.)	rt	sf	-	-	1
Kissiter minimus (Aube.)			1	1	1
Scydmaenidae					
Scydmaenidae gen. & spp. indet.			-	-	1
Orthoperidae					
Corylophus cassidoides (Marsh.)	rt		-	-	1
Staphylinidae					
Megarthrus spp.	rt		-	1	-
Omalium rivulare (Payk.)	rt		-	1	-
Omalium spp.	rt		-	-	1
Coprophilus striatulus F.			1	-	-
Lesteva spp.	oa-d		-	-	1
Oxytelus rugosus (F.)	rt		-	-	1
Oxytelus sculpturatus Grav.	rt		-	-	1
Platystethus arenarius (Fourcr.)	rf		-	1	1
Stenus spp.	u		1	-	-
Xantholinus glabratus (Grav.)			-	-	1
Xantholinus spp.			1	3	1
Othius spp.			-	-	1
Neobisnius spp.			-	1	-
Philonthus spp.			-	2	3
Staphylinus spp.			-	1	-
Lathrobium spp.			1	-	-
Gyrohpnus fracticornis (Mull.)	rt	st	-	1	-
Tachyporus spp.			1	-	-
Aleocharinae gen. & spp. Indet.			-	1	-
Pselaphidae					
Bryaxis spp.			-	1	-
Elateridae					
Adelocera murina (L.)	oa-p		-	-	1
Nitidulidae					
Pria dulcamarae (Scop.)			-	-	1
Cucujidae					
Oryzaephilus surinamensis (L.)	g	ss	-	-	1
Cryptophagidae					
Cryptophagus spp.	rd-h	sf	-	-	1
Atomaria spp.	rd-h	st	-	1	1
?*Atomaria* spp.	rd-h	st	1	-	-
Lathridiidae					
Encimus minutus (L.)	rd-h	st	1	-	-
Encimus spp.	rd-h	st	-	1	1
Anobiidae					
Grynobius planus (F.)	l	sf	-	1	-
Stegobium paniceum (L.)			-	-	1
Anobium punctatum (Geer.)	l	sf	-	-	2
Scarabaeidae					
Trox scaber (L.)	rt	sf	-	1	-
Geotrupes spp.			1	1	1
Onthophagus spp.			1	-	-
Oxymous silvestris (Scop.)	rt	st	-	1	1
Aphodius fimetarius (L.)	oa-rf		-	1	3
Aphodus granarius (L.)	oa-rf		1	1	-
Aphodius spp.	oa-rf		2	1	-
Chrysomelidae					
Chaetocnema spp.	oa		-	2	1
Curculionidae					
Apion aenuem (F.)	oa-p		1	-	-
Apion urticarium (Hbst.)	oa-p		-	-	1
Apion spp.	oa-p		-	1	1
Sitona spp.	oa-p		1	-	1
Cidnorhinus quadrimaculatus (L.)	oa-p		-	2	-
Ceutorhyncus spp.	oa-p		1	-	1

*Ecological codings (*Kenward and Hall 1995)
oa (& ob) - Species which will not breed in human housing.
w- aquatic species.
d- species associated with damp watersides and river banks.
rd- species primarily associated with drier organic matter.
rf - species primarily associated with foul organic matter often dung.
rt - insects associated with decaying organic matter but not belonging to either the rd or rf groups.
g-species associated with grain.
l - species associated with timber.
p – phytophage species often associated with waste areas or grassland and pasture.
h - members of the 'house fauna'. This is a very arbitrary group based on archaeological associations
(Hall and Kenward 1990).

Synathropic codings (Kenward 1997 and pers. comm.).
sf - faculative synanthropes - common in 'natural' habitats but clearly favoured by artificial conditions.
st - typically synanthropes - particularly favoured by artificial habitats but believed to be able to survive in nature in the long term.
ss - strong synanthropes - essentially dependant on human activity for survival.

TABLE 5.21 THE INSECTS RECOVERED FROM THE WELL

from these samples are presented in Table 5.21. The taxonomy follows that of Lucht (1987).

Where applicable, each species of Colepotera has been assigned to one, or more, ecological groupings and these are indicated in the second column of Table 5.21. These groupings are derived from the preliminary classifications outlined by Kenward (1978). The classifications used here replicate that used in Kenward and Hall (1995) where possible. The groupings themselves are described in the key at the end of Table 5.21. The various proportions of these groups, expressed as percentages of the total Coleoptera present in the faunas, are shown in Table 5.22 and Figure 5.28.

Some of the Coleoptera have also been assigned codes based upon the extent of their synanthropy. This includes the proportions of Kenward's (Hall and Kenward 1990; Kenward and Hall 1995) 'house fauna' (species apparently particularly associated with human settlements) and those with known synanthropic preferences. These codes are derived from those used by Kenward (1997). The synanthropic groupings are described at the end of Table 5.21 and the individual codes for the relevant species are shown in Column 3 of Table 5.21. The proportions of these synanthropic groupings, expressed as a percentage of the total fauna is presented in Table 5.23 and Figure 5.29.

sample number	No 4, 4–5m	No 5, 5–6m	No 6, 6–7m
number of individuals	26	32	43
number of species	21	27	38
oa-total	50.0%	34.4%	34.9%
oa-w	7.7%	3.1%	4.7%
oa-d	0.0%	0.0%	2.3%
oa-p	11.5%	9.4%	11.6%
L	0.0%	3.1%	4.7%
Rd	7.7%	6.3%	7.0%
Rf	11.5%	12.5%	11.6%
Rt	7.7%	18.8%	18.6%

TABLE 5.22 THE PROPORTIONS OF ECOLOGICAL GROUPS OF INSECTS RECOVERED FROM THE WELL

	No 4, 4–5m	No 5, 5–6m	No 6, 6–7m
st	7.69%	12.50%	6.98%
sf	7.69%	9.38%	16.28%
ss	0.00%	0.00%	2.33%
h	7.69%	6.25%	4.65%

TABLE 5.23 THE PROPORTIONS OF SYNANTHROPIC GROUPS OF INSECTS RECOVERED FROM THE WEL l

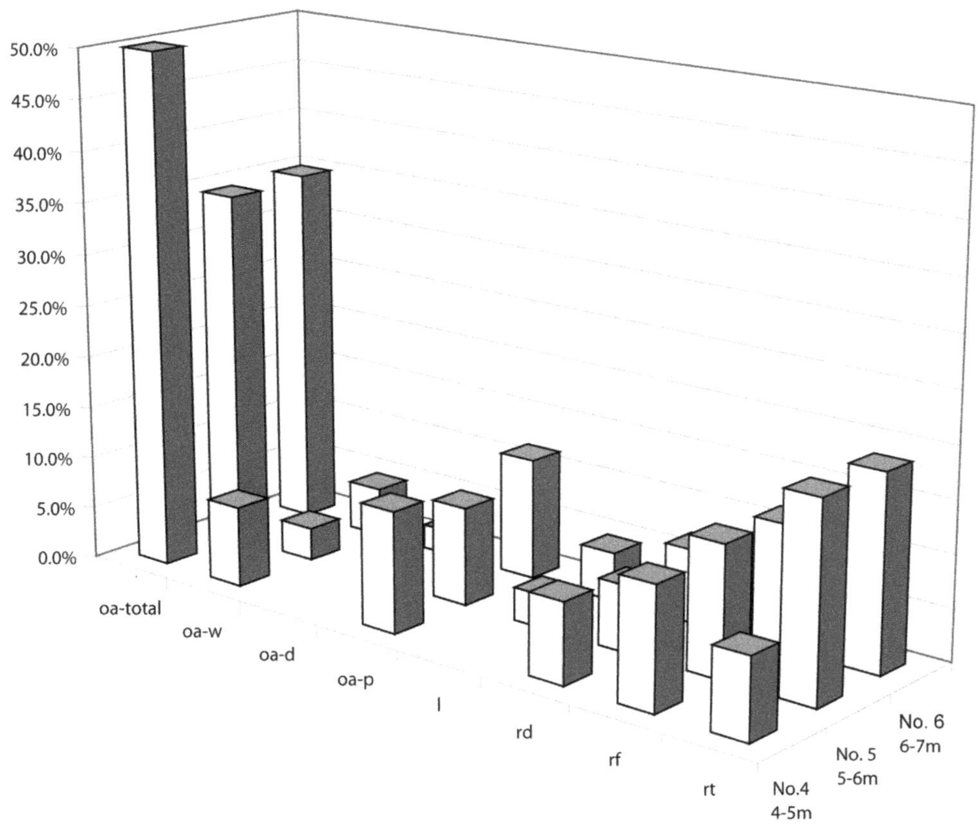

FIGURE 5.28 PROPORTIONS OF THE ECOLOGICAL GROUPS OF INSECTS FROM THE WELL

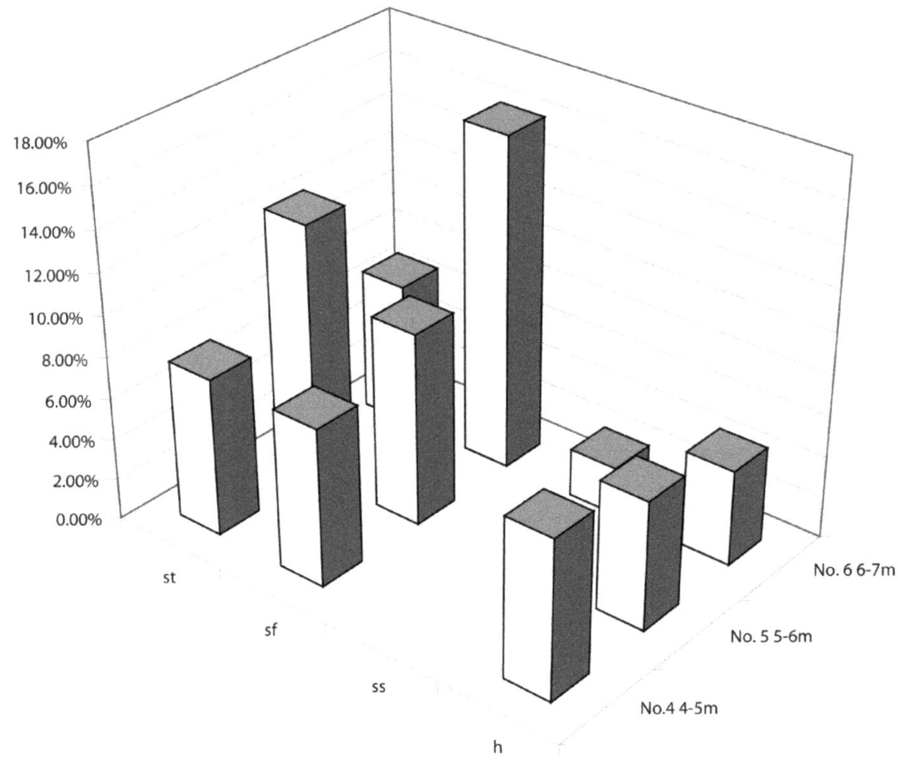

FIGURE 5.29 PROPORTIONS OF SYANTHROPIC GROUPS RECOVERED FROM THE WELL

RESULTS

Only the three lowest samples (Samples 4, 5 and 6) contained substantial remains of insects. It would seem that the upper layers of the well are not adequately waterlogged for preservation of insect remains.

The samples produced three moderately sized insect faunas that are very similar in their nature and the three samples will therefore be discussed together. The majority of the taxa recovered essentially are from external environments (ecological group oa) and account from between 34–50% of the fauna recovered (See Table 5.21 and Fig. 5.28). Typically these are species such as the ground beetles *Harpalus aeneus, Pterostichus melanarius, Calathus fuscipes* and *Synotomus foveatus*, which are all associated with open ground, cultivated fields and wasteland (Lindroth 1974). A similar landscape is suggested by the presence of a number of plant-feeding species of beetle that are associated with open ground and cultivated areas. Stinging nettle (*Urtica dioica* L.) is indicated by both *Apion urticarium* and *Cidnorhinus quadrimaculatus* (Koch 1992). *Apion aenuem* is associated solely with the common mallow (*Malva sylvestris* L.) and *Sitona* with clover (*Trifolium* spp.).

Grassland and pasture may also have been present in the area. This is clearly indicated by *Onthophagus, Geotrupes* and *Aphodius* 'dung beetles' recovered. However, all of these species appear to have a wide flight potential and as a result it is difficult to suggest how distant this environment may have been.

It is also clear that settlement debris and matter must have also entered the well. This is suggested by the relatively high proportions of synanthropic species recovered (Table 5.23 and Fig. 5.29). These are all typically associated with a variety of decaying settlement waste and rubbish. Examples of taxa that indicate this are the various *Cercyon* species, the Histeridae, most of the Staphylinidae, the Cryptophagidae and the Lathridiidae. These are essentially all components of a reoccurring group of species that Kenward (Kenward and Hall 1995) has labelled 'house fauna' and are clearly strong indicators for settlement and housing. This is also the case with *Anobium punctatum,* the domestic woodworm. Also present were two species of grain and food pest. This is the single individual of the 'saw toothed grain beetle' *Oryzaephilus surinamensis* that is associated with spoilt grain and grain storage. *Stegobium paniceum* is the 'bread' or 'drug store' beetle which feeds on a wide range of stored materials in human settlement. Given the comparatively limited range of synanthropic species present, and that this component of the faunas is not strongly dominant, these species have probably entered the well from nearby settlement as part of the general accumulation of settlement waste rather than the deliberate dumping of material.

CONCLUSION

Insect faunas from wells are notoriously difficult to interpret (Hall *et al* 1980). The problem is that wells often act as 'pitfall' traps that accumulate both insects and archaeological material from a wide range of environments. It is often difficult to disentangle whether aspects of the faunas are local to the surroundings of the well, or are from a much wider environment. However, it is clear that at The Butts, in this period, there was probably a cleared landscape with pasture and cultivated land. It would also seem that settlement was probably adjacent, and that settlement rubbish was either discarded into the well or entered it as part of the normal 'fall out' of settlement deposition and activity (Kenward 1978).

A number of well faunas from the Roman period have been analysed. The vast majority of these faunas seem to have formed in a similar way to that suggested for The Butts, with a gradual incorporation and build up of material from the surrounding environment. These include a number of rural Roman sites such as the wells at Piddington, Northhamptonshire (Simpson 2001), Haybridge, Essex (Smith 1997), Farmoor, Oxfordshire (Robinson and Lambrick 1979) and Tiddington, Warwickshire (Robinson unpublished).

Interestingly, Roman wells associated with urban and military sites often appear to derive their insect faunas through another process. This is the deliberate dumping of settlement waste. Clear examples of this are the spoilt grain and settlement waste in the Bedern and Skeldergate wells at York (Hall *et al* 1980; Kenward *et al* 1986), grain waste at Inveresk, East Lothian (Smith 2001) and stable waste at Mill Lane, Mancetter, Warwickshire (Smith 1997).

DISCUSSION

The core of the Roman settlement enclosed by extensive earthwork defences (excavated at Lich Street, Fig. 1.2; Barker 1968–69a, fig. 7) lay to the south of The Butts. This excavation has provided a good outline of land use within a small area of the extramural region of Roman Worcester from the 2nd to 4th century.

The extramural occupation extended to the north and east of The Butts, possibly in the form of a ribbon development, along at least one road to Greensforge and perhaps also along High Street–Foregate Street. There would appear to be some correlation between this pattern of occupation and the Second (Worcester) Terrace geological deposits of the River Severn, on which the historic centre is located. The unfolding history from recent excavations implies that this was not a continuous settlement, but rather more sporadic, with buildings dispersed along the line of the roads, and changing in use and layout throughout the Roman period.

Given the gravel terrace deposits and the proximity of the river at Worcester, it would be reasonable to assume some prehistoric settlement. Barker (1968–69, 13–15) considered that the focus of this lay within the southern half of the historic city, although recent evidence has not been forthcoming. Evidence for Prehistoric occupation comprises only a small amount of lithics and prehistoric pottery from excavations to the north (Kardonia: WCCAS 1995, fig. 3.3 and Dalwood *et al* 1997), and late Bronze Age pottery and a possible Iron Age ditch from excavations at Lich Street (Barker 1968–69a). Unfortunately, excavations at The Butts also failed to provide any further evidence for prehistoric activity, with the earliest features dated to the early 2nd century.

PHASE 1: 2ND AND EARLY 3RD CENTURIES

Features from Phase 1 mostly comprised of truncated pits, short gullies and ditches. While there are no features that can be clearly attributed to the 1st century, the finds from these features date broadly from the late 1st to early 3rd century, with the bulk of the material dating to the 2nd century. The most substantial feature is a large ditch (F107) at the southern extent of Area A. This ditch is aligned roughly east–west and measures approximately 3.7m wide and 0.9m deep. It may have had a parallel fence to the north (stakeholes F110 to F112). A pit (F106) cut into the northern edge of the ditch contained an almost complete 2nd-century *tazza* (incense burner). Whilst these vessels are commonly linked to ritual activity, they have recently been found in association with domestic contexts.

All other Phase 1 activity is notably to the south of the Area A ditch and fence, such that the ditch and fence would seem to mark the northern extent of 2nd- and early 3rd-century activity, and might therefore represent a boundary. Phase 1 activity seems to have focused mainly on pits and gullies, notably in Area B. These are all fairly shallow features, and were probably truncated by ground levelling in the post-medieval period. This seems likely, given the absence of either medieval or ploughsoil horizons, with post-medieval demolition layers directly overlying Roman features. The Phase 1 pits could have been dug as rubbish pits, but their sizes, distribution and fills are perhaps more consistent with small-scale gravel quarrying, followed by their convenient reuse for the disposal of waste. While it is difficult to infer broader economic changes, a decline in this activity towards the end of the 3rd century does appear to coincide with a general contraction of settlement in 4th-century Worcester (Dalwood and Edwards 2004, 18).

A layer of cobbles (F300) in Area D may represent a narrow trackway or yard, aligned roughly east–west and up to 3m in width. Consideration has been given to the presence of a Roman road aligned north–south along the eastern edge of this site, and if the Area D feature was a trackway, then it could have connected with this. However, evidence for this road beyond Blackfriars (Fig. 1.2) is scant and no evidence for a road was recorded in Area E.

Phase 1 contexts produced significant quantities of iron-smelting waste, particularly from Pit 404 in Area E, but unfortunately no remains of the associated furnaces. Similar ironworking waste products also dated to the 2nd to 3rd century were recorded at Deansway, approximately 500m to the south (Dalwood and Evans 2004, 43–44). The similar date, composition and form of the slag at both sites suggests that they were produced using similar raw materials and processes, possibly at the same ironworking centre, with the most likely source of the ore in the Forest of Dean. This further confirms Worcester as a centre for the iron-smelting industry during the 2nd century, but in the absence of furnaces, the role of the northern suburbs in this activity remains unclear.

The charred plant assemblage demonstrates the consistent use of cereal crop processing by-products in the form of spelt chaff. It seems likely that this was being used as fuel and then the charred debris disposed of in ditches and pits. This may also have provided fuel for the iron-smelting furnaces; however, this also represents a repeated pattern consistent with the disposal of domestic rubbish from the 2nd to 4th century. Similar evidence for the use of chaff and cereal crop processing by-products was also recorded at Farrier Street (de Rouffignac 1992), Blackfriars (Moffett 1987) and Deansway (Moffet 2004a).

PHASE 2: 4TH CENTURY

On the basis of the pottery, many of the pits were broadly dated to the 2nd or early 3rd century, and some did produce exclusively 2nd-century forms. In several cases, however, this conflicts with the dating of the small finds, many of which date from the 4th century or later. For example, while the pottery from pit F202E in Trench 2 was broadly dated to the 2nd century, the pit produced a coin (*Antoninianus*) of Tetricus I dated AD 270–73. This may suggest that some of the assemblage was either very old at the time of deposition or, as is more likely, was residual. It is possible that as open space to the north of the Roman town, this area was used for the disposal of waste during the 2nd century. This may explain the high proportion of 2nd-century material from later (Phase 2) pits and ditches, and also implies that the supply of pottery reduced during the later 3rd century, or that waste was dumped elsewhere. The paucity of pottery within Phase 2 may also cast some doubt over the early date for Phase 1 pits from which no small finds were produced.

The most common pottery wares recovered from the pits were from local sources although, not surprisingly, a significant proportion of the assemblage comprised wares produced by major industries from outside the local area. A change in pottery supply between Phases 1 and 2 fits the general pattern at Worcester, with earlier assemblages dominated by local wares. In particular, the supply of Severn Valley wares declined in the 4th century with a corresponding increase in BB1 and Oxfordshire colour-coated wares.

Finds which might be associated with structures were negligible from both Phase 1 and 2, and while later phases may have obliterated any structural features, there really is very little evidence that buildings were ever located on the site during the Roman period. Several of the smaller pits (Area E: F402, F403, F411 and F412) may have been for posts, but the general absence of structural features and paucity of finds associated with buildings, tends to suggest that the area was predominantly open ground throughout the Roman period

A well (F901, Plot 5) at the western edge of the site appears to date to the late 3rd and 4th centuries. The fill of the cut for the well (9000) produced 2nd- and 3rd-century forms, and the fill suggests that the well was backfilled as a single event towards the end of the Roman period. The environmental evidence from the lower 2m of the well seem to confirm the suggestion that much of the area in the immediate vicinity was pasture or waste ground rather than an area of occupation during the 4th century.

During the 4th century, an east–west hollow way formed part of a thoroughfare within the northern extent of Area A. The hollow way was maintained with a stone surface F100 (1006) and appears to have been subject to frequent repairs. Several pits dug to the north of the stone surface (Trench 1) produced late 3rd- and 4th-century pottery forms. Other finds included a 4th-century penannular brooch and a *Follis* of Constantius II (AD 346–50). Much of the area to the south of the road surface probably remained open ground. Here, pits dated to Phase 2 (F400 and F420, area E) were dug much deeper than those from Phase 1, but their function as rubbish dumps appears to be the same. Significant quantities of ironworking waste from Phase 2 pit fills indicate that iron smelting continued to be an important local activity during the 4th century. Interestingly, several furnaces dating to the 3rd and 4th centuries have been discovered at Broad Street 100m to the south of The Butts, and further 3rd- to 4th-century ironworking waste was found at Farrier Street (Dalwood and Edwards 2004, 101).

The stone surface F100 (1006) was probably in use over an extended period. Whilst no pottery was recovered from the surface, one of the thin layers of clay (1012) immediately overlying it at either side produced forms broadly dated to the 3rd and 4th centuries. These layers were not extensive, and were sealed by a much deeper layer (1005) which filled the remainder of the hollow way.

THE HOLLOW WAY FILL (1005) AND POST-ROMAN DEPOSITS

The nature of this layer and the mechanism whereby it accumulated are not certain, but it is thought that the most likely explanation is that the material used to fill the hollow way, as part of post-medieval levelling, was derived from a late Roman midden, dump or cultivation horizon.

At many sites in Worcester a soil layer described as a 'dark earth' often seals the late Roman levels. At Farrier Street the level was at least 1m thick (Dalwood *et al* 1994, fig. 6), and up to 1.2m thick at the Kardonia site (Robin Jackson, pers. comm.). Castle Street to the north and Sansome Street to the east (Darlington 1988) also provided evidence for dark earth deposits, 0.6m in depth in each case. The nature of such soils has been debated (Esmonde Cleary 1989, 147–48). Scientific analyses of the deposits at Farrier Street (Macphail 1994) demonstrated the presence of abundant charcoal, iron slag, daub and animal coprolites, indicative of small-scale industry and household refuse. On the other hand, more recent analysis of the 'dark earth' at Deansway suggests that it formed under grassland pasture from the 4th century, and was reworked during the Anglo-Saxon period by earthworm activity.

Layer 1005 at 14–24 The Butts is not what is traditionally considered a 'dark earth', as it is limited in extent, and essentially cannot be considered the same consistent agricultural layer in evidence at Deansway. Furthermore, the range of finds, including Roman, medieval and post-medieval pottery and artefacts, is much broader than the mid 3rd-century to 10th-century finds from the dark earth at Deansway (Bryant 2004, 329–30: Period 6). The post-Roman finds comprised four sherds of pottery, two medieval 'pens', clay pipe, glass, brick, tile and stone.

The deposit could, perhaps, represent the remains of a *plaggen* soil, 'a cultivated soil on the edge of a settlement enriched by the addition of manure and rubbish' (Limbrey 1975, 335; Darlington and Evans 1992, 33) which formed in the locality of the hollow way. This was the interpretation suggested for sherds in the dark earths at Castle Street (Dalwood *et al* 1997, 7, appendix 4, context 204), which had a comparable average sherd weight of 22g. However, it seems more plausible that the area continued to be used for dumping into the 4th century after which it was turned over to pasture. Post-medieval levelling may then have been responsible for the redeposition of Roman midden deposits within the hollow way. Whilst, therefore, the significant pottery assemblage from 1005 was probably redeposited, it has a high average sherd weight, little abrasion, and on the basis of a wide range of fabrics and forms, can be regarded as a reasonably distinct group. The assemblage has provided an opportunity to examine 4th-century pottery, previously under-represented in Worcester.

PHASE 3: 17TH AND 18TH CENTURIES

The stratigraphic and artefactual evidence for activity from the late Roman to post-medieval periods is scant. The analysis of charred plant and insect remains from the upper layers of the well fill (F901), whilst undated, would also infer that following the Roman period the site became waste ground, rather than any form of occupation or agriculture. The site certainly lay directly to the north of, and outside, the medieval city wall and ditch. While there is a likelihood of garden use with associated small-scale and temporary buildings, later activities would have easily removed any evidence for this (Kenyon 1999, 17). In addition, the presence of late material within the hollow way (1005) would tend to argue against this. It therefore seems likely that the site comprised open fields outside the medieval suburb of Worcester, which is known to have extended along Foregate Street to the east of the site. The rear plots of this suburb may have terminated at Farrier Street (Fig. 1.3).

A ditch (F418) in Area E may have resulted from the redefining of extra-mural Civil War defences located to the north of the former medieval city walls. Such activities occurred in many cities in order to strengthen fortifications during the war. A similar Civil War ditch was recorded at Farrier Street, and could be part of the same activity. Any buildings in the vicinity of The Butts would have been cleared to prevent their use by the enemy and to provide adequate firing zones, although the archaeological evidence from 14–24 The Butts suggests there may have been no substantial structures here to be demolished. As suggested by Jackson (1992), it is possible that the medieval ground surface was wholly truncated, such that Civil War and later clearance removed evidence for medieval use of the site. Even so, the total absence of medieval deposits, particularly given the extent and survival of Roman deposits, does imply that the area remained unoccupied prior to, and during the Civil War.

It is possible that between the end of the Civil War and John Doharty's survey of 1741, the ditch (F418/ F417), became part of a stream channel which fed into a larger channel to the west, shown on Doharty's map as leading north from the city ditch. During the 18th century, the archaeological evidence suggests that the area began to be developed, particularly for industrial buildings, stables, workshops and offices.

PHASE 4: 19TH CENTURY

At some point between the late Roman period and the 19th century the area was heavily scarped, possibly involving the removal of the uppermost layers from across much of the eastern half of the site. This may have been as a response to flooding in order to level-up low-lying or flood-prone areas. A layer of brown silt (4007E, Trench 4) dated to the Roman period, suggests that the western half of the site was not subject to such harsh levelling.

ACKNOWLEDGEMENTS

The archaeological work was commissioned by Phil Brown on behalf of Miller Homes (West Midlands) Ltd. The excavation was supervised by Bob Burrows with the assistance of Ioannis Altsitzoglou, Bob Bracken, Richard Cherrington, Dhaminder Chuhan, Paul Harris, Kristina Krawiec and Sally Radford. The watching brief was undertaken by Helen Martin, Eleanor Ramsey and Erica Macey-Bracken. The evaluation was supervised by Eleanor Ramsey with the assistance of Bob Burrows, Heather Hopkins, Philip Mann, Charlotte Neilson, Andrew Newton, Andrew Rudge and Dan Slater. Trench 4 was excavated by Charlotte Neilson and Eleanor Ramsey. The evaluation and excavation were managed for Birmingham Archaeology by Gary Coates. The illustrations were by Nigel Dodds, Helen Moulden and Bryony Ryder. Thanks also go to James Dinn who monitored the project on behalf of the Worcester City Council and commented on the report. Eleanor Blakelock would like to thank Dr Tim Young who shared his expertise on provenancing ores and assisted with the identification of the ore. Also thanks to Dr Gerry McDonnell for his assistance in identifying the various microstructures present in the Worcester bloom. David Smith is grateful to Dr Kenward for supplying him with a listing of the Colepotera species in each grouping. Felicity Wild would like to thank Brenda Dickinson for drawing her attention to the parallel to decorated sherd no. 3 and for supplying the die number and other comments on the potter's stamp. Copy editing was carried out by Malcolm Hislop.

CHAPTER 6: CONCLUSIONS

Simon Butler

The four projects reported in this volume have examined deposits within archaeologically sensitive areas on the northern and eastern edges of the historic centre of the city of Worcester. At all four sites, the stratigraphy is characterised by Roman and post-medieval deposits, with a distinct lack of intervening material. Ditch features relating to Civil War remodelling of the city's defences were located at St Martin's Gate, 8–12 The Butts, and 14–24 The Butts, and it seems that this and later activity was responsible for the disturbance and removal of earlier material. At 1 The Butts, the creation of the medieval city defences in the 13th century had been responsible for the removal of earlier deposits, but a stone-lined well and other features of Roman date survived on the berm between the medieval city wall and ditch. At all four sites, the Roman deposits yielded a significant array of features and rich assemblages of pottery and other materials, which have added to an understanding of life and industry in the suburbs of Roman Worcester. An archaeological context of research themes and priorities was outlined in the introduction to this volume, and this provides the following framework for evaluating the results from the four projects.

ROAD NETWORK AND SETTLEMENT MORPHOLOGY

One of the research priorities for these excavations was to locate evidence for the road network of Roman Worcester. The three excavations at The Butts were specifically concerned with connecting the previously identified road section to the north at Farrier Street, with road sections identified to the south at Blackfriars and Broad Street. The best candidate for this is an area of gravelling and metalling discovered at 8–12 The Butts. However, the layers are too dispersed to be certain, and the excavator has suggested they may represent a roadside yard or market rather than a main road. Consequently, the course of a main road northwards from the Roman town centre towards Greenforge remains only partially substantiated. Similarly, the site at St Martin's Gate has produced no actual evidence for a main road entering Worcester here on a postulated line from Droitwich.

However, whilst the excavations have not substantiated the main road network itself, they have added to an understanding of the nature, extent and morphology of the Roman settlement. The excavations at St Martin's Gate have shown that the site was in active use during the 1st to 3rd century, extending the known area of Roman settlement along the eastern edge of the Worcester Terrace. Most of the activity here appears to have consisted of the dumping of domestic and industrial waste in pits, but there are indications that the area was also a focus for industrial ironworking, and the presence of some structures is hinted at. The results draw attention to the possible importance of the Frog Brook, as well as the main roads, as a factor influencing settlement morphology. They also draw attention to the possibly diffuse nature of the Roman settlement, with several foci of activity in addition to a 'town centre'.

The Butts excavations have started to give a picture of the general layout of an area in the northern suburb of Roman Worcester, and changes in local land use over time. Thus, during the 2nd century, a ditch was constructed in the northern area of The Butts, possibly with a fence to its north, and this may have acted as a boundary. To the south of this lay an area of largely open ground which seems to have been used initially as a quarry area for sand and gravel extraction, perhaps the source of gravel used in surfacing nearby roads and yards. At least two such gravel surfaces are evidenced on the site for the 2nd century; one from 8–12 The Butts, and one from 14–24 The Butts. The gravel pits were then used as convenient rubbish dumps, whilst the yards and open ground seemed to have served essentially agricultural or industrial functions. This activity continued into the late 3rd and 4th centuries, by which time the area had at least two wells, and a cobbled road ran along the northern boundary on a roughly east–west alignment. These results have added to the developing map of minor roads, yards and enclosures in Roman Worcester, such as those already known from Deansway, and demonstrate the growing morphological complexity of the 2nd- and 3rd-century town.

The regularity of the 2nd-century street pattern at Deansway pointed to a degree of deliberate planning, and it may be that the gravelling and metalling of road and yard surfaces at The Butts should be seen in this context. The activity would have required some organisation and investment, which raises questions about the social and economic dynamics involved. Similarly, the dumping of iron slag and other waste evidenced at The Butts and St Martin's Gate needs to be understood in terms of the individuals or groups responsible for it. The scale of dumping in discrete areas, such as Area B and Trench 3 at 14–24 The Butts, and the associated laying out of yards and minor roads, perhaps points to a co-ordinated policy of waste disposal, industrial expansion and settlement growth by a civic authority. As discussed for Deansway by Dalwood and Edwards (2004, 47), the question then becomes one of whether such developments in Roman Worcester were led by an imperial authority, or by a 'native form of urbanism' controlled by local elites (descendents of late Iron Age clan leaders) who, by the 2nd century, may have been able to develop the small town as a local power base (Millett 1990, 150).

BUILDINGS AND STRUCTURES

Of particular interest is the evidence from the well at 1 The Butts for at least two substantial masonry structures. The actual positions of the buildings are not evidenced, but they are likely to have stood nearby. One is dated to the 2nd century or earlier, while the other is 4th century in date. The later building may have been a public bath house, making use of the local springs which still supplied baths here in the post-medieval period. This evidence makes a substantial contribution to the small body of existing evidence for masonry structures in Roman Worcester, and also indicates that the northern suburb was not exclusively an area of timber-built farms, workshops and fields, but also included more substantial structures. At St Martin's Gate, the presence of further structures is hinted at by an isolated posthole, a possible beam slot and a possible limestone wall, perhaps indicating that the eastern suburb also contained some substantial buildings. However, beyond these discoveries, there is a distinct lack of structural evidence from the excavations, which tends to suggest that these areas of the Roman town were predominantly rural rather than urban in nature throughout the Roman period.

ECONOMY

The excavations have produced some good evidence for the type of activities located in these northern and eastern suburbs. Both agricultural and industrial activities are indicated, giving the impression of mixed land-usage, although there is a lack of structural evidence to confirm any particular emphasis in different areas. Two of the four sites produced significant quantities of iron slag, which are thought to indicate ironworking in their vicinity. Thus, at 14–24 The Butts waste from iron smelting was dumped in rubbish pits from the 2nd century through to the 4th century, and at St Martin's Gate ironworking waste was similarly dumped in rubbish pits, here dated from the 1st through to the 3rd century. The dating of this activity at the two sites is interesting. Thus, a later Roman shift in ironworking away from St Martin's Gate compares well with a similar process in the eastern suburb at Sidbury. But the 2nd-century occurrence of ironworking in the northern suburb at The Butts contrasts with the later 3rd- and 4th-century dates from nearby Farrier Street and Broad Street, and is more in line with the evidence from Deansway further to the south. This makes it difficult to substantiate the shifting pattern of ironworking across the town suggested by Jackson (2004), which pointed to a shift towards the northern suburb only in the later 3rd and 4th centuries, although it must be born in mind that no actual ironworking structures were identified at The Butts.

Of particular interest from St Martin's Gate is possible evidence for iron smithing, from the retrieval there of spheroidal hammerscale. This helps to substantiate the idea that the iron industry in Roman Worcester was concerned not just with iron production, but also with the manufacturing of finished products. In general terms, the evidence of ironworking from The Butts and from St Martin's Gate corroborate the importance of this industry to the economy of the Roman town, and this needs to be seen in the context of developing production industries in the town and its region in the 2nd and 3rd centuries. Dalwood and Edwards (2004, 47) place Worcester's iron industry on a par with Malvern's pottery industry and Droitwich's salt industry, which would suggest that Worcester iron or iron products entered a wide economic network, being distributed extensively across Roman Britain. The 'golden age' of the Roman Empire in the 2nd and 3rd centuries thus seems to be reflected in Worcester by the development of production industries and the (planned) growth of the settlement.

At St Martin's Gate, the occurrence of charcoal in association with iron slag points to the use of charcoal as a fuel for the furnaces, and the predominance of cereal chaff in the charred plant remains at St Martin's Gate and The Butts can also be interpreted as a fuel source for the furnaces. In addition, they indicate that cereal cultivation formed part of the local economy, and that this was primarily concerned with spelt wheat production (*Triticum spelta*). This is consistent with other plant assemblages from Roman Worcester, but the scale of production and its relative role in the economy, and the layout of fields and location of crop-processing activities, are all currently impossible to assess. At 1 The Butts, a possible cultivation soil of early Roman date was observed sealed beneath the medieval city wall, and this could indicate that cultivation plots existed here in the early Roman period. The soil was cut by later Roman features, which might suggest that cultivation was overtaken through time by other activities, and this would be consistent with the apparent settlement growth and industrial expansion of the 2nd century.

Animal bone representation was generally poor at all the sites, and there is no clear evidence to indicate stock rearing, as found at Deansway. However, the predominance of cattle in the animal bone assemblage from 14–24 The Butts might be seen as consistent with stock rearing on the site, or at least consistent with the importance of cattle in the local economy. Interestingly, the author (see Hancox, Chapter 5) draws attention to the predominance of cattle foot bones in The Butts assemblage, and how this could suggest the possibility that tanning was undertaken nearby.

LATE/ POST-ROMAN 'DARK EARTH' DEPOSITS

So-called 'dark earth' deposits are a common feature of sites in Worcester and in other towns and cities in Britain. The deposit characteristically lacks *in situ* features, but is generally also the main source of archaeological material from the late Roman and early post-Roman periods. Such 'dark earths' were recorded overlying the Roman deposits at St Martin's Gate and at 8–12 The Butts, where they range in thickness from 0.2–0.5m, and are dated by pottery inclusions to the 3rd or 4th century. These deposits may, however, have been truncated by later activity, as there is a lack of overlying medieval material and clear indications of post-medieval disturbance. Post-medieval activity seems to have completely removed any 'dark earths' at 1 The Butts, where they were not recorded at all. At 14–24 The Butts, post-medieval activity also seems to have severely disturbed any original 'dark earth' deposit, leading to its removal from some areas of the site and its possible redeposition in others, principally filling the hollow way with a mix of material ranging in date from Roman to post-medieval times. The material filling the hollow way may, however, have derived from a late Roman dump or midden.

The truncated 'dark earths' at St Martin's Gate and 8–12 The Butts indicate that accumulation began in the late 3rd or 4th century, which is consistent with other sites in Worcester. Their lack of stratification and lack of *in situ* features suggests that their accumulation marks a change to some form of lower intensity land use. This has been interpreted at Deansway, with the aid of soil micromorphology, as a change of site use from open settlement to grassland pasture, and is seen as an indication of contraction of settlement at Worcester during the 4th century. The evidence from St Martin's Gate and from 8–12 The Butts might be similarly interpreted, although no soil analysis was undertaken. The evidence from 1 The Butts and 14–24 The Butts is too disturbed to allow much comment, but it is interesting that the east–west aligned hollow way at 14–24 The Butts may have remained in use for some after the 4th century. Also of interest are the stakeholes recorded cut into the 'dark earth' at 8–12 The Butts, which could relate to later Anglo-Saxon or medieval use of the area. However, both sites lie well beyond the Anglo-Saxon *burh* defences, as recorded at Deansway, and also lie just outside the medieval city wall and ditch. A lack of Anglo-Saxon and medieval material at all four sites additionally implies an absence of settlement here in post-Roman times until the 16th century and later, although the apparent truncation of deposits by post-medieval activity will have removed any such evidence.

ABBREVIATIONS

BUFAU	Birmingham University Field Archaeology Unit
HWCCAS	Hereford and Worcester County Council Archaeological Service
HWCM	Hereford and Worcester County Museum
WCC	Worcester City Council
WCCAS	Worcestershire County Council Archaeological Service

BIBLIOGRAPHY

Albarella, U, and Davies, S, 1994 *The Saxon and medieval animal bones excavated from 1985–1989 from West Cotton, Northamptonshire*, Ancient Monuments Laboratory Report 17/ 94

Allen, D, 1998 *Roman Glass in Britain*, Princes Risborough: Shire Publications

Andrew, R, 1984 *A Practical Pollen Guide to the British Flora*, Quat Res Ass, Techn Guide 1

Ayto, E G, 1999 *Clay Tobacco Pipes*, Shire Album 32, Princes Risborough: Shire Publications

Barfield, L H, 2006 Bays Meadow villa, Droitwich: excavations 1967–77, in D Hurst (ed.), *Roman Droitwich: Dodderhill fort, Bays Meadow villa, and roadside settlement*, CBA Res Rep 146, London: Council for British Archaeology, 78–242

Barfield, L H, and Hartley, K H, 2006 Mortaria, in Barfield 2006, 155–57

Barker, P A, 1968–69 The origins of Worcester, *Trans Worcestershire Archaeol Soc*, 3rd Ser, 2, 7–116

Barker, P A, 1968–69 Excavations on the Lich Street development site, 1965–66, in Barker 1968–69, 44–60

Barker, P A, 1968–9b The Broad Street site, in Barker 1968–9, 63–83

Barker, P A, White, R H, Pretty, K B, Bird, H, and Corbishley, M J, 1997 *The Baths Basilica, Wroxeter: excavations 1966–1990*, English Heritage Archaeol Rep 8

Bayley, J, Dungworth, D, and Paynter, S, 2001 *Archaeometallurgy*, English Heritage Centre for Archaeology Guidelines

Beijerinck, W, 1947 *Zadenatlas der Nederlandsche Flora: Ten Behoeve van de Botanie, Palaeontoogie, Bodemcultuur en Warenkennis*, Wageningen: Veenman and Zonen

Beijerinck, W, 1976 *Zadenatlas der Nederlandsche Flora: Ten Behoeve van de Botanie, Palaeontoogie, Bodemcultuur en Warenkennis*, Amsterdam: Backhuys and Meesters

Bennett, K D, 1994 *Annotated Catalogue of Pollen and Pteridophyte Spore Types of the British Isles*, unpublished report

Bishop, M C, 1996 *Finds from Roman Aldborough: a catalogue of small finds from the Romano-British town of Isurium Brigantum*, Oxbow Monograph 65, Oxford: Oxbow

Boardman, S, and Jones, G, 1990 Experiments on the effects of charring on cereal plant components, *J Archaeol Sci* 17, 1–11

Booth, P, 1991 Inter-site comparison between pottery assemblages in Roman Warwickshire: ceramic indicators of site status, *J Roman Pottery Stud* 4, 1–10

Booth, P, and Evans, J, 2001 *Roman Alcester, Northern Extramural Area, 1969–1988 Excavations*, Roman Alcester Series 3, CBA Res Rep127, London: Council for British Archaeology

Booth, P M, and Green, S, 1989 The nature and distribution of certain pink, grog tempered vessels, *J Roman Pottery Stud* 2, 77–84

Bowker, C H, 1982 Environmental research at Wasperton, *West Midlands Archaeol* 25, 25–26

Bowker, C H, 1983 Environmental research at Wasperton: second interim report, *West Midlands Archaeol* 26, 29–35

Bretherton, J, and Pearson E, 1998 *Watching brief at 3–5 The Butts, Worcester*, HWCCAS report

Brown, A, 1994 A Romano-British shell-gritted pottery and tile manufacturing site at Harrold, Bedfordshire, *Bedfordshire Archaeol* 21, 19–107

Bryant, V, 2004 Medieval and early post-medieval pottery, in Dalwood and Edwards 2004, 281–339

Bryant, V, and Evans, C J, 2004 Iron Age and Romano-British pottery, in Dalwood and Edwards 2004, 240–80

Burnham, B C, and Wacher, J, 1990 *The Small Towns of Roman Britain*, London: Batsford

Carreras, C, and Williams, D F, 2003 Spanish olive-oil trade in late Roman Britain: Dressel 23 amphorae from Winchester, *J Roman Pottery Stud* 10, 64–68

Carver, M O H, (ed.) 1980 Medieval Worcester, an archaeological framework, *Trans Worcestershire Archaeol Soc*, 3rd Series, 7, 1–12

Carver, M O H, 1980a The site and settlements at Worcester, in Carver 1980, 15–29

Carson, R A C, and Kent, J P C, 1960 Roman Imperial Coinage of the Later Empire, in *Late Roman Bronze Coinage II AD 324–498*, 41–114, London: Spink

Chirikure, S, and Paynter, S, 2004 Ferrous Waste, in A Lyons *Romano-British Industrial Activity at Snettisham, Norfolk*, East Anglian Archaeology Occasional Papers 18, 42–45

Clapham, A R, Tutin, T G, and Moore D M, 1989 *Flora of the British Isles*, Cambridge: Cambridge University Press

Cleere, H, and Crossley, D, 1995 *The Iron Industry of the Weald,* Cardiff: Merton Priory Press

Coates, G A, and White, R H, 2000 *Final report on an archaeological evaluation on land at 14–20 The Butts Worcester*, BUFAU report 741.01

Cowgill, J, and McDonnell, G, with Gale, R, Mills, J M, Russell, A, and Vince, A, forthcoming, Ironworking technology, in T Ashwin and A Tester (eds), *A Romano-British Settlement in the Waveney Valley: excavations at Scole, 1993–4*, East Anglian Archaeology

Cracknell, S, and Mahany, C (eds) 1994 *Roman Alcester: southern extramural area 1964–1966 excavations, Vol. 1, Part 2: finds and discussion*, Roman Alcester Series, CBA Res Rep 97, London: Council for British Archaeology

Crummy, N, 1983 *The Roman Small Finds from Excavations in Colchester 1971–9,* Colchester Archaeological Report 2, Colchester Archaeological Trust Ltd

Cunliffe, B, 1991 *Iron Age Communities in Britain*, 3rd edn, London: Routledge

Dalwood, C H, and Edwards, R, 2004 *Excavations at Deansway, Worcester 1988–1989: Romano-British small town to late medieval city*, CBA Res Rep 139, London: Council for British Archaeology

Dalwood, C H, Buteux, V, and Darlington, J, 1994 Excavations at Farrier Street and other sites north of the city wall, Worcester, 1988–92, *Trans Worcestershire Archaeol Soc,* 3rd ser, 14, 75–114

Dalwood, C H, Buteux, V A, and Jackson, R A, 1992 Interim report on excavations at Deansway, Worcester 1988–1989, *Trans Worcestershire Archaeol Soc*, 3rd ser, 13, 121–28

Dalwood, H, Buteux, V, and Pearson, E, 1997 *Evaluation at former County Education offices, Castle Street, Worcester*, HWCCAS report

Darling, M J, 2002 The pottery, in J Chadderton (ed.) 2002 *The legionary fortress at Wroxeter: excavations by Graham Webster, 1955–85*, English Heritage Archaeology Report 19, 137–223

Darlington, J, 1988 *Evaluation at Sansome Street, Worcester*, HWCCAS report 14

Darlington, J, and Evans, C J, 1992 Roman Sidbury, Worcester: excavations 1959–1989, *Trans Worcestershire Archaeol Soc*, 3rd ser, 13, 5–104

Davis, S, 1992 *A Rapid Method for recording information about mammal bones from archaeological sites,* Ancient Monuments Laboratory Report 19/ 92

De Rouffignac, C, 1992 The environmental report, in Darlington and Evans 1992, 61

De Rouffignac, C, 1994 Botanical remains, in Dalwood *et al* 1994, 85–86

Dobney, K, Jacques, D, and Irving B, 1996 *Of Butchers and Breeds*, Lincoln Archaeol Stud 5

Dobney, K, Kenward, H, and Roskams, S, 1997 All mixed up and somewhere to go? Confronting residuality in bioarchaeology, in G De Boe and F Verhaegheb (eds), *Method and Theory in Historical Archaeology: papers of the Medieval Europe Brugge 1997 conference*, I A P Rapporten 9, Zellike: Instituut voor het Archeologisch Patrimonium, 81–87

Edwards, R, Griffin, L, and Dalwood, H, 2002 Excavations on the site of the new police station, Castle Street, Worcester, *Trans Worcestershire Archaeol Soc*, 3rd ser, 18, 103–32

Egan, G, and Pritchard, F, 1991 *Dress Accessories 1150–1450*, London: British Museum Publications

Ellis, P, (ed.) 2000 *The Roman Baths and Macellum at Wroxeter*, English Heritage Archaeol Rep 9

Esmonde Cleary, S, 1989 *The Ending of Roman Britain*, London: Batsford

Evans, C J, 2000 The Roman pottery, in R H White, *Excavations in advance of water main relining at Wroxeter Roman city, Shropshire*, BUFAU report 462.02

Evans, C J, 2007 The Roman pottery from Wroxeter and its Hinterland, Whitley Grange, in V L Gaffney, R H White and H Goodchild, *Wroxeter, the Cornovii and the Urban Process: final report on the Wroxeter Hinterland Project 1994–7*, Journal of Roman Archaeology Supplementary Series 68, 146–68

Evans, C J, Jenks, W E, and White, R H, 1999 Romano-British kilns at Meole Brace (Pulley), *Trans Shropshire Archaeol Hist Soc* 74, 1–27

Evans, C J, Jones, L, and Ellis, P, 2000 *Severn Valley Ware Production at Newland Hopfields.: excavation of a Romano-British kiln site at North End Farm, Great Malvern, Worcestershire in 1992 and 1994*, BUFAU Monograph Series 2, BAR British Series 313, Oxford: BAR Publishing

Evans, J, 1994 Discussion of the pottery in the context of Roman Alcester, in Cracknell and Mahany 1994, 144–49

Evans, J, 1999 The end of Roman pottery in the north, in A R Wilmot and P Wilson (eds) *The Late Roman Transition in the North: papers from the Roman archaeology conference, Durham 1999*, BAR British Series 299, 39–46, Oxford: BAR Publishing

Evans, J, Lee, F, and Lindquist, G, 1994 Vessels with incised graffiti, in Cracknell and Mahany 1994, 124–30

Evans, J, and Millett, M, 1992 Residuality revisited, *Oxford J Archaeol* 11(2), 225–40

Fægri, K, and Iversen, J, 1989 *Textbook of Pollen Analysis*, 4th edn, Chichester: Wiley

Featherby, R, 2004 Gresham Street *tazza*, *Study Group for Roman Pottery Newsletter* 38, December 2004. www.sgrp.org.uk.

Figueiral, I, 1992 The Fuels, in M G Fulford and J R L Allen (eds), Iron-making at the Chesters Villa, Woolaston, Gloucestershire: survey and excavation 1987–9, *Britannia* 23, 159–208

Fowler, E, 1960 The origins and development of the penannular brooch in Europe, *Proc Prehist Soc* 26, 149–77

French, A, 2006 Oxford parchment and colour-coated wares, in Barfield 2006, 154

Fulford, M G, and Allen J R L, 1992 Iron-Making at the Chesters Villa, Woolaston, Gloucestershire: survey and excavation 1987–91, *Britannia* 23, 159–215

Gale, R, 1999 Charcoal, in A Fitzpatrick, C A Butterworth and J Grove *Prehistoric and Roman sites in East Devon: the A30 Honiton to Exeter Improvement DBFO Scheme, 1996–9. Vol. 2: Romano-British sites*, Wessex Archaeology report 16.2, 372–84

Gale, R, 1997 Fuel associated with 3rd–4th century metal-working slag at Welwyn Hall, Welwyn, Hertfordshire, unpublished report prepared for Hertfordshire Archaeological Trust

Gale, R, 2000a The fuels, in A J Barber and N Holbrook, A Roman iron-smelting site at Blakeney, Gloucestershire: excavations at Millend Lane 1997, *Trans Bristol Gloucestershire Archaeol Soc* **118**, 33–60, 51–54

Gale, R, 2000b The identification of charcoal from Bridgewater 1963 excavation, in R Jackson, *The Roman settlement at Ariconium, near Western-under-Penyard, Herefordshire; an assessment and synthesis of the evidence*, WCCAS report 833

Galloway, J A, Keene, D, and Murphy, M, 1996 Fuelling the City: Production and Distribution of Firewood and Fuel in London's Region, 1290–1400, *Economic History Review* 69 (3), 447–72

Gillam, J P, 1976 Coarse fumed ware in northern Britain and beyond, *Glasgow Archaeol J* 4, 57–89

Godwin, H, 1956 *The History of the British Flora*, Cambridge: Cambridge University Press

Grant, A, 1982 The use of tooth wear as a guide to the age of domestic ungulates, in B Wilson, C Grigson C and S Payne (eds), *Ageing and Sexing Animal Bones from Archaeological Sites*, BAR Brit Ser 109, 91–108, Oxford: British Archaeological Reports

Greene, S, Dickinson, B, Evans, J, Hancocks, A, Hartley, B, Hartley, K, Pengelly, H, and Williams, D F, 2001 Pottery (Areas 1–6), in A E Jones Roman Birmingham I, Metchley Roman Forts Excavations 1963–4, 1967–9 and 1997, *Trans Birmingham Warwickshire Archaeol Soc* 105, 90–97

Greig, J, 1982 The interpretation of pollen spectra from urban archaeological deposits, in A R Hall and H K Kenwood, *Environmental Archaeology in the Urban Context*, CBA Res Rep 42, London: Council for British Archaeology, 47–65

Greig, J, 1988 The interpretation of some Roman well fills from the Midlands of England, in H Kuster (ed.), *Der prähistorische Mensch und seine Umwelt: Fetschrift für Udelgard Körber-Grohne*, Forschungen und Berichte zur Vor- und Frühgeschichte in Baden-Würtemberg 31, Stuttgart: Konrad Theiss, 367–78

Greig, J, 1990 Practical ecology: experiments in growing traditional cornfield weeds and a comment on their archaeological records in Britain, in D Robinson (ed.) *Experimentation and Reconstruction in Environmental Archaeology*, Symposia of the Association for Environmental Archaeology No. 9, Roskilde, Denmark 1988, Oxford: Oxbow Books, 41–62

Grierson, P, 1975 *Numismatics*, London: Oxford University Press

Griffin, L, 2002 The Pottery, in Edwards *et al* 2002, 103–32

Griffin, L, 2005a The Roman pottery, in T Vaughan 2005 *Archaeological recording on land off Evesham Road, Upper Moor, Pershore, Worcestershire*, WCCAS report 1309, 9–21

Griffin, L, 2005b The Roman pottery, in S Griffin, L Griffin and R Jackson *Salvage recording and evaluation at Throckmorton Airfield, Throckmorton, Worcestershire*, WCCAS, internal report 917

Groves, A W, 1952 *Wartime Investigations into the Hematite and Manganese Ore Resources of Great Britain and Northern Ireland*, Permanent Records of Research and Development Monograph No. 20.703, Mineral Resources Division, Colonial Geological Surveys, Ministry of Supply, London

Hall, A R, and Kenward, H K, 1990 *Environmental Evidence from the Colonia*, The Archaeology of York 14/ 6, London: Council for British Archaeology

Hall, A R, Kenward, H K, and Williams D, 1980 *Environmental Evidence from Roman Deposits in Skeldergate*, The Archaeology of York 14/ 3, London: Council for British Archaeology

Hancocks, A, 2004 Romano-British pottery, in A E Jones, Roman Birmingham 2. Metchley Roman Forts: excavations 1998–2000 and 2002. The eastern and southern annexes and other investigations, *Trans Birmingham Warwickshire Archaeol Soc* 108, 47–59

Hassall, M, and Rhodes, J, 1975 Excavations at New Market Hall, Gloucester, 1966–7, *Trans Bristol Gloucestershire Archaeol Soc* 93, 15–100

Hattatt, R, 1989 *Ancient Brooches and Other Artefacts*, Oxford: Oxbow Books

Hill, P V, 1964 *The Coinage of Septimius Severus and his Family of the Mint of Rome, A.D. 193–217*, London: Spink

Hillman, G C, 1981 Reconstructing Crop husbandry practices from charred plant remains of crops, in R Mercer, *Farming Practice in British Prehistory*, Edinburgh: Edinburgh University Press, 123–62

Hillman, G C 1982a Crop husbandry at the medieval farmstead, Cefn Graenog, in R S Kelly, Excavation of a medieval farmstead at Cefn Graenog Clynnog, Gwynedd, *Bull Board Celtic Stud* 4, 859–908

Hillman, G C, 1982b Evidence for malting spelt, in R Leech, *Excavations at Catsgore 1970–1973, a Romano-British Village*, Western Archaeological Trust Excavation Monograph 2, Gloucester, 137–41

Hillman, G C 1984a Traditional husbandry and processing of archaic cereals in recent times: the operations, products and equipment which might feature in Sumerian texts. Part I: The glume wheats, *Bull Sumerian Agriculture* 1, 114–52

Hillman, G C 1984b Interpretation of archaeological plant remains: the application of ethnographic models from Turkey, in W van Zeist and W A Casparie (eds), *Plants and Ancient Man: Studies in Palaeoethnobotany*, Rotterdam: Balkema, 1–41

Hillman, G C, Mason, S, de Moulins, D, and Nesbitt, M, 1996 Identification of archaeological remains of wheat: the 1992 London workshop, *Circaea* 12 (2), 195–210

Hillson, S, 1992 *Mammal Bone and Teeth: an Introductory Guide to Methods of Identification*. London: Institute of Archaeology, University of London

Houghton, A W J, 1964 A Roman pottery factory near Wroxeter, *Trans Shropshire Archaeol Hist Soc* 57, part 2, 101–11

Howe, M D, Perrin, J R, and Mackreth, D F, 1980 *Roman Pottery from the Nene Valley: a guide*. Peterborough City Museum Occasional Paper 2

Hughes, P M, 1980 Houses and property in post-Reformation Worcester, in Carver 1980, 269–92

Hurst, D, and Rees, H, 1992 Pottery fabrics: a multi-period series for the county of Hereford and Worcester, in S Woodiwiss (ed.), *Iron Age and Roman Salt Production in the Medieval Town of Droitwich: excavations at the Old Bowling Green and Friar Street*, CBA Res Rep 81, 200–209, London: Council for British Archaeology

Ireland, C, 1983 The Roman Pottery, in C Heighway, *The East and North Gates of Gloucester and Associated Sites: excavations 1974–81*, Western Archaeological Trust Excavation Monograph 4, 96–124, Bristol

Jackson, R, 1992 *Evaluation at The Butts, Worcester*, HWCCAS report 106

Jackson, R A, Hurst, J D, and Pearson, E A, 1996 A Romano-British settlement at Leylandii House Farm, Norton and Lenchwick, *Trans Worcestershire Archaeol Soc*, 3rd ser, 15, 63–72

Jackson, R, 2004 Production: Roman ironworking, in Dalwood and Edwards 2004, 100–105

Jacomet, S, 1989 *Prehistoric Cereal Finds: a guide to the identification of prehistoric barley and wheat finds*, a translation by James Greig of *Prähistorische Getreidefunde: Eine Anleitung sur Bestimmung Prähistorischer Gersten- und Weizen-Funde* (1987, Basel: Institut für Botanik)

Jones, G, 1984 Interpretation of archaeological plant remains: ethnographic models from Greece, in W van Zeist and W A Casparie (eds), *Plants and Ancient Man: studies in palaeoethnobotany*, Rotterdam: Balkema, 43–61

Jones, E, and Vyce, D, 2000 *Worcester Magistrates Court, Castle Street, Worcester: an interim statement of an archaeological excavation*, Archaeological Investigations Ltd report

Jones, M K, 1981 The development of crop husbandry, in M Jones and G Dimbleby (eds), *The Environment of Man: the Iron Age to the Anglo-Saxon Period,* BAR Brit Ser 87, Oxford: British Archaeological Reports, 95–127

Jones, M K, 1988a The arable field: A botanical battleground, in M K Jones (ed.) *Archaeology and the Flora of the British Isles*, Oxford University Committee for Archaeology, Oxbow Monograph 14, Oxford: Oxbow Books, 86–92

Jones, M K, 1988b The phytosociology of early arable weed communities with special reference to southern England, in H Küster (ed.), *Der prähistorische Mensch und seine Umwelt*, Forschungen und Berichte zur Vor- und Frühgeschichte in Baden-Württemberg Band 31, Stuttgart: Konrad Theiss Verlag, 43–51

Karnitsch, P, 1971 *Sigillata von Iuvavum (Salzburg): die reliefverzierte Sigillata im Salzburger Museum Augusteum*, Salzburg

Kent, D H, 1992 *List of Vascular Plants of the British Isles*, London: Botanical Society of the British Isles

Kenward, H K, 1978 *The Analysis of Archaeological Insect Assemblages: a new approach*, The Archaeology of York 19/ 1, London: Council for British Archaeology

Kenward, H K, 1997 Synanthropic insects and the size, remoteness and longevity of archaeological occupation sites: applying concepts from biogeography to past 'islands' of human occupation, *Quat Proc* 5, 135–52

Kenward, H K, and Hall, A R, 1995 *Biological Evidence from Anglo-Scandinavian Deposits at 16–22 Coppergate*. The Archaeology of York 14/ 7, London: Council for British Archaeology

Kenward, H K, Hall, A R, and Jones, A K G 1980 A tested set of techniques for the extraction of plant –and animal macrofossils from waterlogged archaeological deposits, *Science and Archaeology* 22, 3–15

Kenward, H K, Hall, A R, and Jones, A K G, 1986 *Environmental Evidence from a Roman Well and Anglian Pits in the Legionary Fortress*, The Archaeology of York 14/ 5, London: council for British Archaeology

Kenyon, D, 1999 *Croft Road/ The Butts and Newport Street, Worcester: archaeological* assessment, Cotswold Archaeological Trust report 991116

Koch, K,1992 *Die Kafer Mitteleuropas*, Okologie Band 3, Krefeld: Goecke and Evers

Kresten, P, and Hjärthner-Holdar, E, 2001 Analyses of the Swedish ancient iron reference slag W-25: R, *Historical Metallurgy* 35 (1), 48–51

Lawson, A J, 1976 Shale and jet objects from Silchester, *Archaeologia* 105, 241–73

Lee, F, Lindquist, G, and Evans, J, 1994 Romano-British Coarse Pottery, in Cracknell and Mahany 1994, 3–92

Letts, J, 1996 Charred plant remains, in R J Williams, P J Hart and A T L Williams (eds), *Wavendon Gate, a late Iron Age and Roman settlement in Milton Keynes*, Buckinghamshire Archaeol Soc Monograph Series 10, 244–56

Limbrey, S, 1975 Soil Science and Archaeology, London: Academic Press

Lindroth, C H, 1974 Coleoptera, Carabidae, in *Handbooks for the Identification of British Insects* 4 (2), London: Royal Entomological Society of London, 1–148

Lucht, W H, 1987 *Die Käfer Mitteleuropas (Katalog)*, Krefeld: Goecke and Evers

MacGregor, A, 1985 *Bone, Antler, Ivory, and Horn: the technology of skeletal materials since the Roman period*, Beckenham: Croom Helm

MacGregor, A, Mainman, A J, and Rogers, N S H, 1999 *The Archaeology of York: Craft, Industry and Everyday Life: bone, antler, ivory and horn from Anglo-Scandinavian and medieval York*, The Archaeology of York 17/ 12, London: Council for British Archaeology

Mackreth, D F, 1992 The Brooches, in J Darlington and J Evans, Roman Sidbury, Worcester: Excavations 1959–1989, *Trans Worcestershire Archaeol Soc*, 3rd Ser, 13, 73–77

Mackreth, D F, 1994 Copper Alloy and Iron Brooches, in Cracknell and Mahany 1994, 162–70

Macphail, R I, 1994 Soil micromorphology, in Dalwood *et al* 1994, 83–84

Macphail, R I, 2004 Soil micromorphology, in Dalwood and Edwards 2004, 558–67

McDonnell, G, 1995 *Iron working Processes*, Historical Metallurgical Society Archaeology Datasheet 3

McDonnell, G, and Swiss, A, 2004 Ironworking residues, in Dalwood and Edwards 2004, 148–57

Manning, W H, 1985 *Catalogue of the Romano-British Iron Tools, Fittings and Weapons in the British Museum*, London: British Museum Publications

Marsh, G, 1978 Early second century fine wares in the London area, in P Arthur and G Marsh, *Early fine wares in Roman Britain*, BAR Brit Ser 57, Oxford: British Archaeological Reports, 119–223

Mikkelsen, P H, 1997 Straw in slag-pit furnaces, in L Nørbach (ed.), *Early Iron Production – Archaeology, Technology and Experiments*, Technical Report 3, Lejre: Historical Archaeological Experimental Centre, 63–66

Mikkelsen, P H, 2003 Slag – with an impression of agricultural practices, *Acta Jutlandica* 76(2), 43–48

Millett, M, 1990 *The Romanisation of Britain: an essay in archaeological interpretation*, Cambridge: Cambridge University Press

Millward, J, 2005 *A report on the Roman pottery assemblage from 14–24 The Butts, Worcester: Area E*, Unpubl MA project report, University of Birmingham

Mitchell, A, 1974 *A Field Guide to the Trees of Britain and Northern* Europe, London: Collins

Moffett, L M, 1986 *Crops and crop processing in a Romano-British village at Tiddington: the evidence from the charred plant remains*, Ancient Monuments Laboratory Report 15/ 86

Moffett, L M, 1987 *Two Roman pits from Worcester, Blackfriars,* Ancient Monuments Laboratory Report 203/ 87

Moffett, L M, 1995 *Botanical remains from Worcester Deansway*, Ancient Monuments Laboratory Report 39/ 95

Moffett, L, 2004a Cultivation and site environment, in Dalwood and Edwards 2004, 91–94

Moffett, L, 2004b Botanical remains, in Dalwood and Edwards 2004, 537–56

Moffett, L M, and Ciaraldi, M, 1999 The charred plant remains, in S Palmer (ed.), Archaeological excavations in the Arrow Valley, Warwickshire, *Trans Birmingham Warwickshire Archaeol Soc* 103, 162–70

Monckton, A, 1999a *Charred plant remains from corn driers and other contexts of a Romano-British settlement site at Billesley Manor Farm, Warwickshire*, Ancient Monuments Laboratory Report 25/ 99

Monckton, A, 1999b *Charred plant remains from a Romano-British farmstead at Glebe Farm, Bubbenhall, Warwickshire*, Ancient Monuments Laboratory Report 24/ 99

Monckton, A, 1999c The Environmental Evidence, in A Connor and R Buckley (eds), *Roman and Medieval Occupation in Causeway Lane, Leicester*, Leicester Archaeological Monograph 5, 309–62

Morton, G R, and Wingrove J, 1969 Constitution of bloomery slags. Part 1: Roman, *Journal of the Iron and Steel Institute* 207, 1556–64

Mould, Q, with Webster, G, and Lloyd-Morgan, G, 2000 The Small Finds, in Ellis 2000, 108–44

Mundy, C F, 1986 *Worcester archaeology project 1985/ 6*, HWCCAS interim report

Mundy, C F, 1994 *Watching brief on an extension to Spring Gardens Health Centre*, SMR 100246

Murphy, P, 1989 Plant remains, in J Plouviez, A Romano-British Pottery Kiln at Stowmarket, *Proc Suffolk Inst Archaeol* 37, 8–10

Murphy, P, 1997 *Charred macrobotanical and other remains from a Roman Pottery kiln, Heath Farm, Postwick: an assessment 1997*, unpubl typescript

Murray, E, 2001 The Animal Bone, in G Coates and M Duncan, *An archaeological evaluation on the site of The Green, Forebridge, Stafford,* BUFAU report 757.01, 10–13

Napthan, M, 2002 *Archaeological salvage recording at Tramps, Angel Place, Worcester, WCM100899* Mike Napthan Archaeology, unpubl report

Napthan, M, 2004 *Archaeological works at 1 The Butts, Worcester, WCM 101107, 101108, 101109*, Mike Napthan Archaeology, unpubl report

Neal, D, Wardle, A, and Hunn, J, 1990 *Excavation of the Iron Age, Roman and Medieval Settlement at Gorhambury, St Albans,* English Heritage Archaeol Rep 14

Nesbitt, M, and Samuel, D, 1996 From staple crop to extinction? The archaeology and history of the hulled wheats, in S Padulosi, K Hammer and J Heller (eds) *Hulled Wheats* (*Promoting the Conservation and Use of Underutilized and Neglected Crop 4*), Proceedings of the First International Workshop on Hulled Wheats, 21–22 July 1995 at Castelvecchio Pascoli, Tuscany, Italy, Rome: International Plant Genetic Resources Institute, 41–100

Nye, S, and Jones, M, 1994 Plant remains from the mausoleum site, in R Williams (ed.) *Bancroft: the late Bronze Age and Iron Age settlements and Roman temple-mausoleum. Vol. 2*, Buckinghamshire Archaeol Soc Monograph Series 7, 562–65

Oswald, F, 1936–7 *Index of Figure Types on Terra Sigillata ('Samian Ware')*, University of Liverpool Annals of Archaeology and Anthropology Supplement

Paynter, S, 2002 *The Metalworking Remains from Westhawk Farm, near Ashford, Kent*, English Heritage Centre for Archaeology report 84/ 2002

Paynter, S, 2006 Regional variations in bloomery iron-smelting slag of the Iron Age and Romano-British periods, *Archaeometry* 48, 271–92

Peacock, D P S, 1965–67 Romano-British pottery production in the Malvern district of Worcestershire, *Trans Worcestershire Archaeol Soc*, 3rd Ser, **1**, 15–28

Peacock, D P S, and Williams, D F, 1986 *Amphorae and the Roman Economy: an introductory guide*, London: Longman

Pelling, R, 2000 Charred plant remains, in A Mudd and P Booth (eds) Site of former Hockley Chemical Works, Stratford Road, Alcester: excavations 1994, *Trans Birmingham Warwickshire Archaeol Soc* 104, 53–56

Philpott, R A, 1999 A Romano-British brooch type from North-Western and Northern England, *Britannia* 30, 274–86

Plouviez, J, nd, unpublished, *Roman fort and small town at Pakenham, Suffolk: excavations on the route of the Ixworth by-pass in 1985*, HER ref PKM 005

Price, J, and Cottam, S, 1998 *Romano-British Glass Vessels: a handbook*, CBA Practical Handbook in Archaeology 14

Rackham, O, 1990 *Trees and Woodland in the British Landscape*, London: Dent

Rackham, D J, Gale, R, Holden, T, Izard, K, and Locker, A, 1998 *Parnell Road and Lefrevre Walk, Bow, London, PRB 95 and LEK 95,* Environmental archaeology report for Preconstruct Archaeology

Rawes, B, 1982 Gloucester Severn Valley ware: a study of the Roman pottery forms, *Trans Bristol Gloucester Archaeol Soc* 100, 33–46

Rees, H, 1992 Pottery, in S Woodiwiss (ed.) *Iron Age and Roman salt production and the medieval town of Droitwich.* CBA Res Rep 81, 35–61

Rees, H, 2004 Iron Age and Roman Pottery, in F McAvoy, Dodderhill, Droitwich: excavations 1977–85, in Hurst 2004, 20–30

Rees, H, forthcoming, The Roman pottery, in J Wills forthcoming, *Excavations at Beckford, Worcestershire, 1972–9*

Reynolds, S, 2005 *A report on the Roman pottery assemblage from 14–24 The Butts, Worcester: Area D*, unpubl MA project report, University of Birmingham

Robertson, A S, 1971 *Roman imperial coins in the Hunter coin cabinet, Vol. II. Trajan to Commodus*, London: Oxford University Press

Robertson, A S, 1982 *Roman imperial coins in the Hunter coin cabinet, Vol. V. Diocletian (Reform) to Zeno*, London: Oxford University Press

Robinson, M, and Lambrick, G, 1979 *Iron Age and Roman Riverside Settlements at Farmoor, Oxfordshire*, CBA Res Rep 32

Robinson, M A, nd *Waterlogged insect remains from the three Roman wells at Tiddington, Warwickshire*, unpubl report

Rodriguez-Almeida, E, 1989 *Los Tituli Picti de las Anforas Olearias de la Baetica*, Madrid

Rogers, G B, 1974 *Poteries Sigillées de la Gaule Centrale I: les motifs non figurés,* Gallia Supplément XXVIII, Paris

Samuels, L E, 1980 *Optical Microscopy of Carbon Steels*, Ohio: American Society for Metals

Sawle, J, 1977 *Worcester Sidbury Excavation – The Second Season: background*, HWCM 117 unpubl report

Schmid, E, 1972 *Atlas of Animal Bones for Prehistorians, Archaeologists and Quaternary Geologists*, New York: Elsevier

Seager Smith, R, and Davies, S M, 1993 Black Burnished ware and other southern British coarsewares, in P J Woodward, S M Davies and A H Graham, *Excavations at the Old Methodist Chapel and Greyhound Yard, Dorchester 1981–1984*, Dorset Nat Hist Archaeol Soc Monograph Series 12, 229–89

Simpson, T, 2001 The Roman well at Piddington, Northhamptonshire, England: an investigation of the Coleopterous Fauna, *Environmental Archaeol* 6, 91–96

Smith, D N, 1997 *The insect remains from Mancetter Mill Lane, Roman Well*, unpubl report prepared for Keith Scot, director of excavations at Mancetter

Smith, D N, 2001 *The insects from Inverskgate well*, University of Birmingham Environmental Archaeology Services report 30

Stace, C, 1991 *New Flora of the British Isles*, Cambridge: Cambridge University Press

Stace, C, 1997 *New Flora of the British Isles*, 2nd edn, Cambridge: Cambridge University Press

Stanfield, J A, and Simpson, G, 1958, *Central Gaulish Potters*, London: Oxford University Press

Starley, D, 1995, *Hammerscale*, Historical Metallurgy Society Archaeology Datasheet 10

Sutherland, C H V, 1974 *Roman Coins*, London: Barrie and Jenkins

Symonds, R P, 1992 *Rhenish wares: fine dark coloured pottery from Gaul and Germany*, Oxford University Committee for Archaeology Monograph 23

Symonds, R P, 1997 The Roman pottery, in P Barker, R White, K Pretty, H Bird and M Corbishley *The Baths Basilica Wroxeter: excavations 1966–90*, English Heritage Archaeol Rep 8, 269–318

Timby, J, Evans, J, Darling, M J, and Faiers, J E, 2000 The Pottery, in Ellis 2000, 193–312

Timby, J, 2004 Later prehistoric and Romano-British pottery from Childswickham, Worcestershire, in C Patrick and D Hurst *Archaeological survey and excavation along the Cotswold Spring Supply Trunk Main: archive report*, WCCAS internal report 1140, 16–38

Tomber, R S, 1980 *A petrological assessment of Severn Valley ware: kilns and selected distribution*, unpubl MSc dissertation, University of Southampton

Tomber, R S, 1985 Pottery, in A R Wilmot and S P Q Rahtz, An Iron Age and Roman settlement outside Kenchester (Magnis), Herefordshire: excavations 1977–9, *Trans Woolhope Natur Field Club* 45, 99–145, fiche frames 1–59

Tomber, R S, 1991 Methods for investigating deposit homogeneity, *J Roman Pottery Stud* 4, 59–68

Tomber, R, and Dore, J, 1998 *The National Roman Fabric Reference Collection: a handbook*, Museum of London Archaeology Service Monograph 2

Tomlin, R S O, and Hassall, M W C, 2006 Roman Britain in 2005. III. Insciptions. B: Instrumentum Domesticum, *Britannia* 37, 474–85

Tutin, T G, Heywood, V H, Burgess, N A, Moore, D, Valentine, D H, Walters, S M, and Webb, D A, 1964–80, *Flora Europaea,* 5 volumes, Cambridge: Cambridge University Press

van der Veen, M, 1989 Charred grain assemblages from Roman-period corn driers in Britain, *Archaeol J* 146, 302–19

van der Veen, M, 1992 *Crop Husbandry Regimes*, Sheffield Archaeology Monograph 3

van der Veen, M, 1999 The economic value of chaff and straw in arid and temperate zones, *Vegetation History and Archaeobotany* 8, 211–24

VCH, 1924 *Victoria History of the County of Worcester,* Vol. 4

Von den Driesch, A, 1976 *A Guide to the Measurement of Animal Bones from Archaeological Sites*, Peabody Museum Bulletin 1, Harvard University

Walker, I, 1959 Excavations on a Romano-British site at Astley, 1956–58, *Trans Worcestershire Archaeol Soc* 35, 29–57

Webster, P V, 1976 Severn Valley Ware: a preliminary study, *Trans Bristol Gloucestershire Archaeol Soc* 94, 18–46.

Webster, P V, 1993 Coarse pottery, in P J Casey, J L Davies and J Evans *Excavations at Segontium (Caernarfon) Roman Fort, 1975–79*, CBA Res Rep 90, 250–308

Webster, PV, 1996 *Roman Samian Pottery in Britain*, CBA Practical Handbook in Archaeology 13

Wheeler, R E M, and Wheeler, T V, 1932 *Report on the Excavation of the Prehistoric, Roman and Post-Roman site at Lydney, Gloucestershire*, Rep Res Comm Soc Antiq London 9

White, R H, 1988 *Roman and Celtic Objects in Anglo-Saxon Graves*, BAR Brit Ser 191, Oxford: British Archaeological Reports

White, R, and Baker, N, 2000 *A desk-based assessment of the Star Hotel, Worcester and adjacent land*, BUFAU report 672

Whitworth, A, and Edwards, R, 2001 *Evaluation at Tallow Hill and Pheasant Street, Worcester*, WCCAS report 881

Williams, D F, 1977 The Romano-British Black-Burnished industry: an essay on characterization by heavy mineral analysis, in D P S Peacock (ed.) *Pottery and Early Commerce: characterization and trade in Roman and later ceramics*, 163–220

Williams, D F, and Peacock, D P S, 1983 The importation of olive-oil into Roman Britain, in J M Blazquez and J Remesal (eds) *Produccion Y comercio del Aceite en la Antiquedad*, 2nd International Conference at Seville, Madrid: Universidad Complutense, 263–80

Willis, S, 2000 The Iron Age and Roman pottery, in R Jackson *The Roman settlement at Ariconium, near Weston-under-Penyard, Herefordshire: an assessment and synthesis of the evidence*, WCCAS report 833

Willis, SH 2004 Samian, in A E Jones, Roman Birmingham 2: Metchley Roman forts: excavations 1998–2000, and 2002, *Trans Birmingham Warwickshire Archaeol Soc*, 108, 60–64 and 107

Willis, S H, 2005, *Samian Pottery, a resource for the study of Roman Britain and beyond: the results of the English Heritage funded Samian project*, an e-monograph in *Internet Archaeology* 17

Woodward, A, and Leach, P J, 1993 *The Uley Shrine*, English Heritage Archaeol Rep 17

WCC 2007 *Worcester urban archaeological strategy: an outline resource assessment and research framework for the archaeology of Worcester*, Worcester City Council

WCCAS 1995 *Kardonia, Farrier Street, Worcester: assessment and post-excavation* proposal, HWCCAS report 22105

Young, C J, 2000 *The Roman Pottery Industry of the Oxford Region*, BAR Brit Ser 43 (reprint of 1977 edition with new introduction and updated bibliography)

Young, T P, and Thomas, G R, 1999 Provenancing iron ore from the Bristol Channel Orefield: the cargo of the medieval Magor Pill boat, in A M Pollard (ed.) *Geoarchaeology: exploration, environments,* resources, Geological Society Special Publication 165, 103–21

Zohary, D, and Hopf, M, 1994 *Domestication of Plants in the Old World*, Oxford: Clarendon Press

Zohary, D, and Hopf, M, 2000 Domestication of Plants in the Old World: the origin and spread of cultivated plants in West Asia, Europe, and the Nile Valley, 3rd edn, Oxford: Clarendon Press

www.ingramcontent.com/pod-product-compliance
Lightning Source LLC
LaVergne TN
LVHW071218210926
848482LV00015B/840

* 9 7 8 1 4 0 7 3 0 8 0 0 5 *